The Social Organisation of Healthcare Work

Sociology of Health and Illness Monograph Series

Edited by Hannah Bradby
Department of Sociology
University of Warwick
Coventry
CV4 7AL
UK

Current titles:

- **The Social Organisation of Healthcare Work (2006),**
  edited by *Davina Allen and Alison Pilnick*
- **Social Movements in Health (2005),**
  edited by *Phil Brown and Stephen Zavestoski*
- **Health and the Media (2004),**
  edited by *Clive Seale*
- **Partners in Health, Partners in Crime: Exploring the boundaries of criminology and sociology of health and illness (2003),**
  edited by *Stefan Timmermans and Jonathan Gabe*
- **Rationing: Constructed Realities and Professional Practices (2002),**
  edited by *David Hughes and Donald Light*
- **Rethinking the Sociology of Mental Health (2000),**
  edited by *Joan Busfield*
- **Sociological Perspectives on the New Genetics (1999),**
  edited by *Peter Conrad and Jonathan Gabe*
- **The Sociology of Health Inequalities (1998),**
  edited by *Mel Bartley, David Blane and George Davey Smith*
- **The Sociology of Medical Science (1997),**
  edited by *Mary Ann Elston*
- **Health and the Sociology of Emotion (1996),**
  edited by *Veronica James and Jonathan Gabe*
- **Medicine, Health and Risk (1995),**
  edited by *Jonathan Gabe*

Forthcoming titles:

- **The View From Here: Bioethics and the Social Sciences (2006),**
  edited by *Raymond de Vries, Leigh Turner, Kristina Orfali and Charles Bosk*
- **Ethnicity, Health and Healthcare: Understanding Diversity, Tackling Disadvantage (2007),** edited by *Waqar Ahmad and Hannah Bradby*

The Social Organisation of Healthcare Work

Edited by

Davina Allen and Alison Pilnick

Blackwell
Publishing

First published as a special issue of Sociology of Health and Illness Vol. 27 No. 6

BLACKWELL PUBLISHING
350 Main Street, Malden, MA 02148-5020, USA
9600 Garsington Road, Oxford OX4 2DQ, UK
550 Swanston Street, Carlton, Victoria 3053, Australia

First published 2006 by Blackwell Publishing Ltd

Library of Congress Cataloging-in-Publication Data has been applied for

ISBN 1-4051-3334-1

A catalogue record for this title is available from the British Library.

Set by Graphicraft Limited, Hong Kong
Printed and bound in the United Kingdom
by TJ International, Padstow, Cornwall

The publisher's policy is to use permanent paper from mills that operate a sustainable
forestry policy, and which has been manufactured from pulp processed using acid-free
and elementary chlorine-free practices. Furthermore, the publisher ensures that the text
paper and cover board used have met acceptable environmental accreditation standards.

For further information on
Blackwell Publishing, visit our website:
www.blackwellpublishing.com

Contents

Chapter 1

Making connections: healthcare as a case study in the social organisation of work

Davina Allen and Alison Pilnick

Introduction

The division of labour has been a central preoccupation since the earliest days of sociology, and considered key to an understanding of wider societal structures. Both Durkheim and Marx shared Saint-Simon's basic assumption that *the* most fundamental aspect of human existence was the absolute necessity to produce the means of subsistence (Lee and Newby 1983). They recognised that the starting point for understanding a given society was the way in which it organised the totality of activity necessary to existence, although they interpreted this in different ways. Weber, unlike Durkheim and Marx, did not produce over-arching social theories, but for him, too, the social organisation of work was a key concern. These early works set the agenda for, and provided a stimulus to, successive generations of scholars. Healthcare work has emerged as a particularly popular case for the study of the world of work and occupations. In fact, we would argue that the modern healthcare system is *the* most exciting setting in which to study work and organisations. It is a dynamic, technologically-rich environment, which incorporates interfaces between the public, private and voluntary sectors and entails a complex division of labour comprising professions, occupations, unwaged caregivers, managers and technicians. As such, it provides a natural laboratory for the exploration of many classic sociological problems and the identification of new lines of analysis.

In recent years widespread concerns over cost-containment on both sides of the Atlantic have led to the introduction of new modes of organisational governance in healthcare systems. These have had a profound impact on the social organisation of work, changing the sites of provision, increasing labour intensification, challenging traditional lines of demarcation and prompting the development of new roles and modes of working. Such changes have been buttressed by new ideologies of citizenship which are redefining orthodox professional client relationships and explicitly acknowledging the contribution of lay health work. In addition, the introduction of new medical technologies has changed the working environment for many healthcare workers, leading to shifts in the division of labour and changes to working practice. This has presented new challenges to the development of skills and, in some instances, has been seen as an attack on professional autonomy and clinical judgement. Given the rapidity of change in the healthcare sector, this monograph provides an opportunity to produce an international snap-shot of the current state of play. The chapters here describe different dimensions of current health trends in the UK, North and South America, Australia, Canada, Finland and the Netherlands. Bringing them together in this collection highlights convergence and divergence across national and international contexts, and points to new research directions and developments.

There is a strong tradition of studies of health service organisation and delivery within medical sociology. In a recent review undertaken for the silver anniversary special issue of *Sociology of Health and Illness*, Griffiths (2003) argues that a concern with the social organisation of healthcare could be said to be implicit in every paper ever published in the journal. Whilst acknowledging the value of individual contributors, she observes that papers in this area 'sometimes fail to build on earlier research' but argues that 'when read together they offer a valuable picture of a complex range of healthcare settings and their social organisation' (2003: 155). A further strength of devoting a Monograph to this topic is that it enables such connections to be made more readily and facilitates the establishment of links with developments within the wider field of sociological endeavour. Whilst the roots of the issues considered in the chapters included in this volume can be traced to the classic concerns of the sociology of the division of labour, the authors have augmented their contribution to medical sociology by looking to cognate areas of study such as organisational sociology, sociology of scientific knowledge and studies of technically-mediated collaborative work. Our aim in this introduction is to trace some of these connections.

The division of labour

Despite their very different theoretical destinations, both Durkheim and Marx take as the starting point for their analyses the totality of activities that are necessary in a particular society. At the macro-level, such an approach recognises the importance of all forms of work, whether it is paid or unpaid, carried out in the private, public or voluntary sector (Freidson 1978). It focuses attention on how various occupations form, the organisation and structure of their work, how this division of labour changes over time and the principal drivers for such change. At the micro-level, it highlights the need to study the relationship between the bundles of activities that constitute the technical division of labour and the social roles which comprise the moral division of labour in a system of work. Whilst the particular contributions of occupations themselves may often be presented in essentialist terms, every occupation has a history which may be described in terms of changes in bundles of activities, in the values given them and in the total system of which the occupation is a part (Hughes 1984).

Beyond its obvious applicability to the complex division of labour that characterises the paid work of service providers, the adoption of this kind of theoretical lens is particularly pertinent to the healthcare context where the patient and significant others are participants in the work of maintaining health and managing illness. Medical sociologists have drawn attention to self-care activities and expertise developed through the experience of living with a chronic disease or disability, and some have made the case for inclusion of the client in the division of labour as essential for understanding healthcare work (Stacey 1981, Strauss *et al.* 1985). Work in this area has also directed attention to the invisible caring work that takes place in the domestic sphere. This is particularly pertinent to an understanding of healthcare where the division of labour in the family has been translated into and reinforced through the division of labour in the public domain (Stacey 1981, Gamarnikow 1978, 1991). The gendered nature of caring work is seen by many as a key to understanding the relative status of nursing and other carers in the health services division of labour, the historical shifts in the sites of care between the private and the public sphere, and the associated redistribution of work and redefinition of activities that this has occasioned.

Work as social activity

In contrast to the utilitarian approach of economics, a further contribution of sociology to the study of work is the emphasis which is placed on its inherently social character. Durkheim pointed optimistically to the function of the division of labour in creating social solidarity, whilst Marx underlined the alienated qualities of work under capitalist systems of production, and Weber warned of the dispiriting effects of rationalisation processes on the meaning of work and bureaucratic forms of control constraining the free will of the individual. All in their different ways identify the division of labour as the most powerful mobiliser of persons and draw attention to the importance of work for the creation of collective and individual identities.

The healthcare system has been a particularly fruitful site for the exploration of the activities and culture of a wide range of occupational groups and the processes through which individuals are enrolled into such communities of practice (Becker *et al.* 1961, Olesen and Whittaker 1968, Mumford 1970, Dingwall 1977, Atkinson 1981, Melia 1987, Haas and Shaffir 1987). Membership of such social groupings provides a point of reference from which individuals make meaning of their work and manage identity, although occupations vary in the strength and duration of such an association. Hughes' writing on the 'drama of work' has been particularly valuable in signalling the centrality of social roles in an overall division of labour. He coined the term 'dirty work' to draw attention to the existence of work which threatens occupational identities. There is ample evidence of the applicability of this idea in the healthcare context, with several classic studies of medical sociology linking to this theme (Emerson and Pollner 1976, Jeffrey 1979, Dingwall and Murray 1983, Mizrahi 1986, Brown 1989, Allen 2001). For Hughes, work was a central criterion by which people were judged and by which they judged themselves. He argues that even in the lowest occupations people develop collective pretensions to give their work and thus themselves value in the eyes of each other and outsiders.

Organisms and ecology

An enduring metaphor that has shaped studies of the social organisation of work which has been particularly influential in the study of healthcare is that of the biological organism. Its influence can be traced back to Durkheim, who looked to Darwinian theory to develop an understanding of the evolution and development of the division of labour. In developing this line of thought, Durkheim was heavily influenced by Spencer, and Spencer's understanding of the basic laws of human social organisation (see Turner 1984). Durkheim draws on the observations of Darwin that the struggle between organisms is a function of their similarity. The chances of conflict diminish as the species become more distinct from each other. According to Durkheim, people submit to the same law. In the same city different occupations can co-exist without being obliged mutually to destroy one another because they pursue different objects. The closer functions come to one another, he argues, then the greater the risk of conflict. Orthodox accounts of Durkheim's work have tended to downplay the importance he placed on competition as a driver for change in systems of work, yet a careful reading reveals striking similarities between his writings and those of the Chicago School in which the world of work was treated as an analogue to the City's urban ecology (Dingwall 1983).

The most influential recent exposition of social ecological theory is Abbott's *The System of Professions* (1988). Clearly located in fieldwork study and heavily influenced by the work of Hughes' students (Freidson, Bucher, Strauss), his work is a self-conscious attempt to develop a framework which addresses what he perceives to be the main weaknesses in the sociology of the professions: their failure to study the basic conditions and control of the content of work. Placing interprofessional battles at the heart of his work, Abbott is concerned with the evolution and inter-relationships of professions and the ways in which they control their jurisdiction[1]. Given that their fortunes are interdependent, a profession's success reflects as much the situations of its competitors and the system structure as it does its own efforts. Abbott draws attention to the larger social forces that impact on the system of work. Most consequential for the creation of new tasks and the destruction of others are changes in technology and organisation, but he also points to the effects of changing social values which require the creation of new forms of legitimacy. Abbott parts company with early sociologists of the division of labour, in that for him history is not a simple pattern of trends and development but a complex mass of contingent forces of social ecology. His concern with legitimacy, however, links his work to the parallel development of the new institutionalism in the sociology of organisations (Powell and DiMaggio 1991). This approach places considerable emphasis on the way in which organisations are required to adopt certain forms and processes less as a matter of technical rationality or increased efficiency than as a means of meeting the expectations of significant actors in the environment. Ecological imperatives are as much cultural as economic. Arguably, for example, most health professionals and the organisations in which they work are no more or less attentive to patients than they ever were; they are however now required to document and demonstrate this in elaborate ways in order to be treated as legitimate by governments, insurers, courts and other members of the societal networks in which they are embedded. A failure to comply with these cultural expectations has economic consequences in the erosion of resources.

In one sense, a potential limitation of Abbott's thesis for the study of healthcare work is its restriction to the system of professions, rather than work of all forms[2]. Nonetheless, the contemporary system of healthcare represents an interesting case through which to explore his central thesis. As Freidson (1970) has shown us, health service provision has been defined to a considerable extent by the 'professional dominance' of medicine which historically has exerted significant control over this system of work. The evolution and workplace construction of the boundary with medical practitioners and other occupational groups has been a popular focus of concern within medical sociology, and, given the dynamic nature of the division of labour within health care settings, is likely to remain so. Indeed we are now witnessing the emergence of new areas of sociological interest with the development of international health policies designed to make inroads into medical autonomy. Abbott argues that no profession that fails to deliver a service can stand indefinitely against competitors. As state intervention into the healthcare arena on both sides of the Atlantic increases, the stage is set for testing how far a profession can continue to deliver a service in the face of decreasing legitimacy.

Professions, the state and individuals

The relationship between the professions, the state and citizens in shaping the social organisation of work in society was central to Durkheim's thinking (Durkheim 1957). He envisaged a special role for professions as secondary groupings which mediated between the state and the individual. According to Durkheim '. . . it is out of this conflict of social forces that individual liberties are born' (1957: 63). The working out of such tensions in different contexts has been a key concern in the study of health care work ranging from Parson's classic work on the sick role as a mechanism of social control through to more recent work on the rise of surveillance medicine (*e.g.* Armstrong 2005, Samson 2005). This theme has recently received attention with the introduction of new systems of organisational governance in response to the rising health costs and concerns about service quality. The use of national service frameworks, best practice guidelines and evidence-based practice are being promoted as a means for enhancing provision. Such developments present some fundamental challenges to the traditional ecology of knowledge in healthcare. At one level, these developments raise the spectre of Weber's so-called 'iron cage'[3] of bureaucracy and systems of hyper-rationality. At another, recent studies reported in this collection reveal there is ample opportunity for health professionals to resist such efforts to control and standardise their work.

Division of labour as social interaction

One of Hughes' legacies most evident in studies of healthcare systems is the attention he drew to systems of work at the micro-level. For him the division of labour implied interaction; he regarded it to be a poor term for the differentiation of function in social life as a whole, because it emphasised division rather than integration. For this reason, he did not study any one occupation, as he was more interested in their inter-relationships. Hughes' writings are characteristically fragmented, however, and the clearest articulation of this overall approach came from Freidson (1976). Criticising Smith and Durkheim for paying little attention to the concrete substance of the concept of the division of labour, Freidson argues that the ultimate reality of the division of labour lies in the social interaction of its participants.

> Among the individuals on the factory shop floor, or on the hospital ward, and among the groups engaged in negotiating legislation and formal plans for controlling work, there are boundaries set on what will be considered legitimate to negotiate, how the negotiation will take place, and what bargains can be struck. Some are unselfconscious and taken for granted, some are specifically understood as scientifically incontrovertible necessity, and others as legitimate laws, rules and regulations, and practices . . . [I]n the everyday world from which we abstract conceptions of the division of labor, it seems accurate to see the division of labor as a process of social interaction in the course of which the participants are continuously engaged in attempting to define, establish, maintain and renew the tasks they perform and the relationship with others which their tasks presuppose (Freidson 1976: 311).

One of the attractions of healthcare as a site for the study of work is its characteristically complex division of labour. Traditionally, hospital wards and multidisciplinary team meetings (Griffiths 1998, 2001) have proved to be rich locales for the examination of the division of labour in action. It is perhaps significant that the negotiated order perspective arose from a workplace study of a psychiatric hospital (Strauss *et al.* 1964). Studies of this kind drew attention to the situated character of work roles and the mismatch between formal organisational plans and workplace jurisdictions (Abbott 1988) arising from the need to respond to context-specific contingencies. Health-care settings revealed themselves to be an arena for the playing out of inter- and intra-occupational differences and have yielded fascinating materials through which to study occupational identity and boundary work.

Healthcare providers are increasingly expected to co-ordinate their contri-bution with that of the patient and their unwaged caregivers, who bring different kinds of knowledge and expertise to the interaction, and challenge professional power (Pilnick 1998, Allen 2000). New technology has also had a profound impact on organisational conduct in healthcare work, meaning that as well as human/human interaction, human/machine interaction is a burgeoning field of study. Studies in the field of computer-supported collab-orative work have important implications not only for our understanding of new technologies themselves, but also for the ways in which they are actually used in practice (Luff, Hindmarsh and Heath 2000).

One of the distinctive features of healthcare work is that it is people work and this is highly consequential for its social organisation and practice. As we have seen, the patient/carer might legitimately be considered as a co-worker in the division of labour, but they are also the subject of the healthcare providers' attentions. As Hughes argued in 1956, the line between service and disservice is a fine one. 'In many of the things which people do for one another, the *for* can be changed to *to* by a slight overdoing or a shift in mood' (Hughes 1984: 305). There is now an established body of work in medical sociology which has examined the features of the professional encounter and the processes through which power and control are negotiated (Sharrock 1979, Maynard 1991, Heath 1992, ten Have 1991).

Making connections – from classical sociological theory to the chapters in this volume

Having laid out some of the key concerns for sociological theorists in the field of work and organisations, our second aim in the introduction to this Monograph is to connect these with the specific issues addressed in the chapters presented here. Whilst the continued relevance of these classic sociological concerns is evidenced by their presence in each of the contributions to the Monograph, in their application to the contemporary healthcare systems and their interaction with cognate disciplines they coalesce under a different set of themes.

Ecological approaches to the study of healthcare: integrating macro/micro/ meso perspectives
Interdependence is a key thread in classic sociological analyses, alongside the importance of linking entities with the wider system of which they are a

part. This is a theme which emerges very strongly from the chapters included in this collection, demonstrating the continued influence of this line of thinking. Griffiths (2003) argues that the strengths of some of the work previously published in *Sociology of Health and Illness* lies in its attempts to integrate macro-perspectives on policy and the role of the state with micro-perspectives on the working practices of individuals and groups. Specifically, she highlights the need for sociologists to make linkages between macro-, meso- and micro-levels of study, whilst at the same time calling for researchers to treat the distinction between these levels as a phenomenon interactionally accomplished by organisational members and the topic of research.

In different ways, the three chapters in the first section of this monograph all respond to this call, highlighting both the importance of studying macro-/meso-/micro- relationships in order to understand both the whole picture in terms of healthcare organisation and delivery of services, and the ways in which what is experienced as components of these different levels is subject to interpretation and negotiation. Drawing variously on the examples of managerialist, market-led and public-policy-driven healthcare initiatives, these chapters all examine the ways in which larger social forces impact on the system of work. Echoing Abbott's earlier arguments, they demonstrate that not only developments in organisation and available technologies, but also changes in social values, have a profound effect on healthcare delivery.

Murray and Elston's study of obstetric practice in Chile illuminates the relationships between the macro-level of political decisions, the meso-level of the organisations through which government reforms were enacted, and the micro-level of clinical practice. Their work also underlines the need to follow a process through time to assess the implications for delivery, organisation and outcome of services, thus identifying the way in which planned change is actually experienced. Additionally, this piece serves as a powerful reminder of the importance of identifying the system effects of healthcare reforms and how these might lead to unintended consequences – in this case practitioner efforts to rationalise their workload result in increased medical interventions in childbirth in the form of planned caesarean sections and assisted deliveries.

The importance of making these links is further stressed by Benoit and colleagues in the second chapter included here. In their comparison of the organisation of midwifery services across four countries they note that looking only at macro-level differences cannot fully explain variations in healthcare practice cross-nationally. They highlight the relationship between professions, the state and citizenship, and the non-essentialist nature of occupations. Meso-level differences have tended to be less studied by sociologists of health and illness (with notable exceptions such as Hunter 1979, Griffiths and Hughes 2000) but these have a huge impact on whether services are framed in an individual or a collective manner – and this in turn has an impact on both how they are received and how easy they are to change from the point of view of patient groups, consumer organisations, etc. This chapter provides a clear empirical demonstration of Abbott's argument, as Benoit and colleagues demonstrate the way in which welfare state approaches to the legalisation of midwifery, professional boundaries in the maternity care domain, and consumer mobilisation in support of midwifery and maternity issues are critically interlinked in the emergence of change.

Germov's study of policy reform in Australia provides the third example in this volume. This chapter draws on a neo-Weberian framework of hyper-rationality to explore the impact of managerialism on the work organisation of public sector healthcare professionals. Germov seeks to address the question of whether there is an underlying logic to the way managerialism has affected professional bureaucracies, particularly in terms of how professional autonomy is governed. The chapter provides a further reminder that if we are to understand workplace change in any meaningful way, it needs to be situated within wider organisational, political and economic contexts as well as the experiences of individual actors. However, it also identifies that the standardisation of professional practice is partly self-imposed, arguing that this is best conceptualised as a survival strategy.

Professional autonomy and the 'standardisation' of clinical practice

Though perhaps most evident in Germov's chapter, as indicated above, a second theme that runs to some degree through all three chapters in the first section is the way in which constraints upon the professional autonomy of healthcare professionals come to be experienced. This may be through managerial strategies, financial arrangements, etc., but what is of great significance is the way in which members of some occupational groups (*e.g.* the Chilean obstetricians interviewed by Murray and Elston) can find ways of adjusting to these constraints without significant loss. This theme is addressed more explicitly in the second group of chapters included in this monograph. Of course, an increasingly common way in which constraints upon professional autonomy are applied is through the deployment of new tools and technologies. As Heath *et al.* (2003) note, these tools and technologies have an important impact on the ways in which we work and how we work with others. The organisation of healthcare work has undergone sustained change over the last 20 or 30 years, with much of this change driven by the introduction of new technologies and, particularly, computerisation. These new tools and technologies, however, are not limited to scientifically sophisticated pieces of equipment designed to perform medically complex tasks. The growing use of treatment protocols, managed care procedures and care pathways – in themselves simply paper documents – also have a fundamental impact on the ways in which resources are allocated, professionals practice and lines of accountability are established. Ultimately, healthcare is delivered within particular organisational circumstances and this influences, and may constrain, the way in which illness is identified, managed and treated. Heath *et al.*'s review highlights the need for sociological work to 'address how practical circumstances, procedure convention, culture and the socio-political environment shape, even create, the sense and significance of systems and artefacts' (2003: 79).

Three chapters in this monograph explore variously the ways in which growing attempts to 'standardise' healthcare procedures can be called into question, from a practical as well as a sociological point of view. They all, to a greater or lesser extent, draw on work which has its roots in the sociology of scientific knowledge, in exploring the ways in which the supposed neutrality of particular tools or devices can be challenged. Timmermans and Berg (2003) highlight the way in which the sociology of health and illness has moved over the years from a view that medical technologies provided healthcare providers with effective tools to coerce others into medically-approved lifestyles (*e.g.* Conrad's 1979 work considering psychopharmaceuticals) – a view they categorise as technological deterministic thinking – to a social constructivist perspective that draws heavily on science and technology studies. They suggest that the important message of this latter approach is the dynamic relationship between technology and society: 'social interests shape technologies throughout their design process and usage, and in turn technologies shape the activation of different social constituencies' (Timmermans and Berg 2003: 103).

In this volume, Pinder and colleagues' chapter examining care pathways is a carefully illustrated example of the ways in which these kinds of devices highlight some aspects of patient or illness experience whilst silencing others. At the same time, there is an aspect which is less often the central focus of science and technology studies but is key to understanding the social organisation of healthcare: that pathways, protocols, etc. play a role in defining and redefining occupational boundaries. From this chapter, there are also links to be made to the wider sociological critique of objectivity which is a cornerstone of science and technology studies – that many of these tools can be seen as ideological constructs that reflect and sustain power relationships in medical care and medical work as much as any objective 'truth' about what is the best or most effective treatment/way of working. Pinder and colleagues go so far as to suggest that, if we are to view the production of protocols, pathways, etc. as a process of knowledge construction, then sociologists have a moral responsibility to think about how these processes might be used positively to enable a more just distribution of resources. A more modest goal, however, is to focus on how pathway or protocol activity, through its increase in regulation and control of healthcare practices, also seeks directly or indirectly to regulate and control practitioners.

The second chapter in this section, by Catherine Will, examines the history of tools developed to assess an individual's risk of coronary heart disease. In so doing, it echoes the key concerns raised by Pinder and colleagues, that the continual search to specify and objectify always highlights some aspects and silences others, and that as sociologists we have a duty to look into these silences. The coming together of political, economic and professional agendas elucidated by Will here also links back to the first theme discussed above – the importance, and fruitfulness, of taking an ecological approach to the study of healthcare. In addition, Will describes how standardised tools and approaches can both facilitate and reflect a move towards collective medical autonomy as opposed to individual professional autonomy. Running alongside this is often a shift to collective patienthood, as shown here through the transformation of 'individual risk' of heart disease into the 'global danger' of CHD.

The third and final chapter in this section, by Greatbatch and colleagues, considers similar issues from a different perspective. In their study of the systems and protocols of NHS Direct (a 24 hour nurse-led telephone advice service run by the NHS in the UK) in action, Greatbatch and colleagues examine the ways in which 'standardised' tools can be practically subverted to allow healthcare professionals to use their own expertise. This links with the way in which Timmermans (1998) uses the work of Strauss *et al.* (1985) and Hughes (1958) in the field of organisational studies, as well as Latour's work in the social study of science and technology, to examine the relationship between technology, practice and practical circumstance. Examining the work of resuscitation teams, he demonstrates that the 'potential and power of a technological device to shape an interaction is not pre-given but is realised in practice' (Timmermans 1998: 148).

Greatbatch and colleagues also demonstrate how 'expert systems' can never capture the entirety of a group's occupational and professional tasks, since they generally seek a single answer to complex issues. This argument also runs through the other two chapters included in this section, though less explicitly – can clinical and medical expertise ever be embodied in abstract, universal rules? And if not, given that the ideologies of healthcare professionals generally mitigate against standardisation processes, should we conceptualise variations as reflecting individualised service rather than as a failure of the system or of the individual to follow it?

To a lesser extent, the three chapters discussed immediately above also touch on the notion of identities. Heath *et al.* (2003: 81) identify preserving identities as a prevailing issue within the sociology of health and illness. They trace this theme back to Henderson (1935) and Parsons (1951), in terms of the tension that arises between the formal application of specialist medical knowledge and the practical, contingent and organisational circumstances in which healthcare is delivered and received. There is a widespread view that technologies and tools inevitably depersonalise the experience of health-care from the point of view of the recipient. However, Greatbatch and col-leagues' study is a nice example of the way in which particular tools only gain their ultimate character in and through the ways in which they are used by participants. The authors' methods draw on work examining call-taking in other areas of sociology, most notably ethnomethodologically informed workplace studies (Whalen and Zimmerman 1987, Whalen 1995, Garcia and Parmer 1999) that demonstrate in meticulous detail the way in which inter-action and technology use are bound together. So, though the questions on the NHS direct computer system used by the nurses under study are standardised both in phrasing and order, the way in which they are delivered as they actually use the system serves to repersonalise them.

Timmermans and Berg (2003) suggest that the true challenge for sociologists studying medical technologies is how to conduct research that is relevant to the users and designers of new medical tools and technologies while maintaining a sociological identity. And, furthermore, that in areas such as computer-supported co-operative work this is already being negotiated, in terms of, for example, why clinical practice guidelines are not being used or what modifications beyond the technological might be necessary for a particular piece of machinery to be accepted by home care patients. Studies such as those by Pinder *et al.*, Greatbatch *et al.* and Will begin to give us an insight into this from within the sociology of health and illness, demonstrating how tools and technologies both shape and are shaped by practice.

Changing jurisdictions

The two chapters in the third and final section included in this volume focus on the consequences of shifts in the division of labour for workplace jurisdictions. Stacey's data from the US provide clear examples of the way in which the ever growing social care workforce finds itself taking on roles which were once the remit of professional healthcare workers. Now, however, these tasks have been redefined, and in their redefinition they have passed out of the arena of health and into social care with attendant consequences for their social value. Whilst Stacey's study is US-based, similar redefinitions have occurred on the other side of the Atlantic, and, as Stacey's work reveals, such transformations blur the boundary between waged and unwaged work, where employees are transformed into companions and the work of caring for someone in their own home becomes a labour of love. This chapter reveals that the interpersonal bonds that develop in such circumstances obscure the fundamentally exploitative nature of the relationship in which home care workers are frequently required to undertake work outside their jurisdiction. In one sense, the chapter has links with Murray and Elston's work in that it illustrates that whilst higher status workers may be able to adjust to needs for flexibility in order to control and manage their work, this is arguably not an option for lower-status workers. Given these constraints, how do workers find identity and attach meaning to a job that on the face of it looks thankless either in financial or personal terms? Stacey argues that dignity has a practical function in these circumstances, and this argument might be extrapolated to the range of other health and social care jobs which are similarly badly paid and demanding. This chapter reveals that 'boundary work' is not solely relevant to the more generally considered picture of higher-status professionals trying to keep lower-status ones from encroaching on their territory, as evidenced in the claims of Stacey's study participants that they possess skills that hurried and out-of-touch medical professionals do not. To what extent the social care workers' accounts represent another example of occupational identity work or whether their assertions do have some basis in the practical realities of the social organisation of work (Simmel 1950, Mechanic 1961, Scheff 1961, Stein 1967) is a question that will no doubt continue to occupy sections of the sociological community. However, this chapter is a clear call to examine worker experience rather than simply focusing on theoretical models or concepts such as exploitation, and is a useful reminder of Hughes' insistence on the similarities of the problems of work and in the value of researchers not becoming locked into their sub-disciplines.

The final chapter in the monograph, Goodwin and colleagues' UK-based examination of participation in anaesthesia by individuals both of different occupations and of different statuses within those occupations, is an interesting counterpart to Stacey's work. Goodwin *et al.* draw on Anspach's (1993) work about how different forms of knowledge might be prioritised in identifying why and how some interventions by members of the anaesthetic 'team' are not seen as appropriate; and this chimes with Stacey's data suggesting that home care workers are not seen as appropriate sources of knowledge by other professionals. The key difference for Goodwin and colleagues' anaesthetic nurses and operating department personnel is that because they work as part of a real time team they can potentially gain legitimacy to participate or practice in particular activities over time, by demonstrating their ability to do so. At the same time both chapters demonstrate boundary blurring as a response to organisational difficulties, though they give different answers to the question of when knowledge may legitimately be taken as the basis for action. Goodwin *et al.* also highlight, more than any of the other chapters in this volume, the way in which movement outside the boundaries can jeopardise or threaten another's identity. This has echoes with Fox's (1994) study of surgical work, where he considers how technology contributes to claims of authority and responsibilities, both in terms of warranting physical presence and the right or ability to interpret certain kinds of findings.

Conclusion

The social organisation of healthcare work is a well-established academic field and, given the dynamic and rapidly changing characteristics of contemporary healthcare systems, is likely to remain a popular focus of study. As the chapters considered in this volume illustrate, the development of new technologies, evolving health policies and the emergence of new areas of social concern continue to interact in different contexts to produce a rich vein of issues to capture the sociological imaginations of successive generations of scholars. Recent trends on the international stage have created a new set of topics for social scientific study some of which are touched upon in this Monograph, but others are ripe for capture. As we have seen, new technology has had a profound impact on organisational conduct in healthcare work, meaning that as well as human/human interaction, human/machine interaction is a burgeoning field of study. Studies in the field of computer-supported co-operative work have important implications not only for our understanding of new technologies themselves, but also the ways in which they are actually used in practice which is increasingly likely to be in domestic as well as institutional contexts. The reorientation of modern healthcare systems around a community-based public health model also looks set to raise new research questions as the division of caring work is redrawn across organisational and agency boundaries, and new definitions of citizenship emerge redefining public engagement with the management of health and illness and the design of services. There is scope here for medical sociologists to build on the wealth of empirical studies on lay health behaviours and professional-patient interaction to explore negotiation of these new divisions of labour and knowledge in this changed policy context. Trends towards the democratisation of healthcare, coupled with the burgeoning evidence-based practice agenda, have generated the potential for the evolution of new ecologies of knowledge through which health decisions will be mediated. This will create new challenges for health providers who are increasingly likely to be faced with the requirement to reconcile the reductionist recommendations of best practice protocols with the increasing expectation of user involvement and patient-centred care.

Beyond this, our aspiration would be for members of the medical sociology community to continue to build on the strong tradition of empirical work on the social organisation of healthcare work, to develop theoretical and conceptual models which have broader applicability beyond the healthcare context. As we have seen, some of the most influential theoretical contributions to the sociology of work and occupations have their empirical roots in research in healthcare (Strauss 1964, Freidson 1970), and yet as Davies (2003) observes, in recent years the fields of medical sociology and organisational studies have become increasingly disconnected. As we move into the 21st century the strengthening of such links is now long overdue, and we hope that this volume may make a small contribution towards the re-establishment of such connections.

Acknowledgements

We would like to acknowledge the contribution of Robert Dingwall and Jon Gabe who made helpful comments on an earlier draft of the introduction. We are grateful to Carolyn Wiener who contributed to the promotion of the monograph and abstract selection, but unfortunately had to withdraw from the editorship. Thanks are also due to all those who participated in the peer review process.

Notes

1   There are clear intellectual antecedents with Hughes' concepts of licence and mandate.
2   Abbott observes in note 10 p. 387 (Abbott 1988) that Eliot Freidson, Robert Dingwall and Jim Davis and others have raised the question as to whether he has

not given a general theory of the division of labour. However, Abbott maintains that his focus on abstractions limits the application of this study to expert systems.

3 According to Baehr (2001) the 'iron cage' metaphor comes from Parsons' decisions as translator of 'The Protestant Ethic', although the literal translation would be 'a shell hard as steel'. This is significant because Parsons is rendering Weber's argument as one where an external discipline is imposed on individuals, whereas Weber sees individuals retreating into their private spheres from a world whose routinisation has become intolerable.

References

Abbott, A. (1988) *The System of Professions: an Essay on the Division of Expert Labor*. Chicago: The University of Chicago Press.

Allen, D. (2000) Negotiating the role of expert carers on an adult hospital ward, *Sociology of Health and Illness*, 22, 2, 149–71.

Allen, D. (2001) *The Changing Shape of Nursing Practice: the Role of Nurses in the Hospital Division of Labour*. London: Routledge.

Anspach, R. (1993) *Deciding who Lives: Fateful Choices in the Intensive Care Nursery*. Los Angeles: University of California Press.

Armstrong, D. (2005) The rise of surveillance medicine. In Annandale, E., Elston, M.A. and Prior, L. (eds) *Medical Work, Medical Knowledge and Health Care: a Sociology of Health and Illness Reader*. Oxford: Blackwell.

Atkinson, P. (1981) *The Clinical Experience: the Construction and Reconstruction of Clinical Reality*. Farnborough, Hants: Gower.

Baehr, P. (2001) The 'iron cage' and the 'shell as hard as steel': Parsons, Weber and the Stahlahrtes Gehause metaphor in *The Protestant Ethic and the Spirit of Capitalism, History and Theory*, 40, 153.

Becker, A.S., Geer, B., Hughes, E.C. and Strauss, A.L. (1961) *Boys in White: Student Culture in Medical School*. Chicago: University of Chicago Press.

Brown, P. (1989) Psychiatric dirty work revisited: conflicts in servicing nonpsychiatric agencies, *Journal of Contemporary Ethnography*, 8, 2, 182–201.

Conrad, P. (1979) Types of medical control, *Sociology of Health and Illness*, 1, 1, 1–11.

Davies, C. (2003) Some of our concepts are missing: reflections on the absence of a sociology of organisations in Sociology of Health and Illness, *Sociology of Health and Illness*, 25 (*Silver Anniversary Issue*), 172–90.

Dingwall, R. (1983) Introduction. In Dingwall, R. and Lewis, P. (eds) *The Sociology of Professions: Lawyers, Doctors and Others*. London: Macmillan.

Dingwall, R. (1977) *The Social Organisation of Health Visitor Training*. London: Croom Helm.

Dingwall, R. and Murray, T. (1983) Categorization in accident departments: 'good' patients, 'bad' patients and 'children', *Sociology of Health and Illness*, 5, 2, 127–47.

Durkehim, E. (1933) *The Division of Labour in Society*. London: Collier-Macmillan Ltd.

Durkheim, E. (1957) *Professional Ethics and Civic Morals* (Translated by Cornelia Brookfield). London: Routledge and Kegan Paul.

Emerson, R. and Pollner, M. (1976) Dirty work designations: their features and consequences in a psychiatric setting, *Social Problems*, 23, 243–54.

Fox, N. (1994) Anaesthetist, the discourse on patient fitness and the organisation of surgery, *Sociology of Health and Illness*, 16, 1, 1–18.

Freidson, E. (1970) *Professional Dominance*. New York: Atherton Press Inc.

Freidson, E. (1976) The division of labour as social interaction, *Social Problems*, 23, 304–13.

Freidson, E. (1978) The official construction of work: an essay on the practical epistemology of occupations. Paper presented at the Ninth World Congress of Sociology, Uppsala.

Gamarnikow, E. (1978) Sexual division of labour: the case of nursing. In Kuhn, A. and Wolpe, A.M. (eds) *Feminism and Materialism: Women and Modes of Production*. London: Routledge and Kegan Paul.

Gamarnikow, E. (1991) Nurse or woman: gender and professionalism in reformed nursing 1860–1923. In Holden, P. and Littleworth, J. (eds) *Anthropology and Nursing*. London: Routledge.

Garcia, A.C. and Parmer, P.A. (1999) Misplaced mistrust: the collaborative construction of doubt in 911 emergency calls, *Symbolic Interaction*, 22, 4, 297–324.

Griffiths, L. (1998) Humour as resistance to professional dominance in community mental health teams, *Sociology of Health and Illness*, 20, 6, 874–95.

Griffiths, L. (2001) Categorising to exclude: the discursive construction of cases in community mental health teams, *Sociology of Health and Illness*, 23, 5, 678–700.

Griffiths, L. (2003) Making connections: studies of the social organisation of healthcare, *Sociology of Health and Illness*, 25 (Silver Anniversary Issue), 155–71.

Griffiths, L. and Hughes, D. (2000) Talking contracts and talking care: managers and professionals in the NHS internal market, *Social Science and Medicine*, 51, 209–22.

Haas, J. and Shaffir, W. (1987) *Becoming Doctors: the Adoption of the Cloak of Competence*. London: JAI Press.

ten Have, P. (1991) Talk and institution: a reconsideration of the 'asymmetry' of doctor-patient interaction. In Boden, D. and Zimmerman, D. (eds) *Talk and Social Structure: Studies in Ethnomethodology and Conversation Analysis*. Cambridge: Polity Press.

Heath, C. (1992) The delivery and reception of diagnosis in the general practice consultation. In Drew, P. and Heritage, J. (eds) *Talk at Work: Interaction in Institutional Settings*. Cambridge: Cambridge University Press.

Heath, C., Luff, P. and Sanchez Svensson, M. (2003) Technology and medical practice, *Sociology of Health and Illness*, 25 (Silver Anniversary Issue), 75–96.

Henderson, L.J. (1935) Physician and patient as a social system, *New England Journal of Medicine*, 212, 2, 819–23.

Hughes, E. C. (1958) *Men and their Work*. Glencoe: Free Press.

Hughes, E. (1984) *The Sociological Eye*. New Brunswick and London: Transaction Books.

Hunter, D. (1979) Coping with uncertainty: decisions and resources within health authorities, *Sociology of Health and Illness*, 1, 1, 40–68.

Jeffrey, R. (1979) Normal rubbish: deviant patients in casualty departments, *Sociology of Health and Illness*, 1, 1, 90–108.

Lee, D. and Newby, H. (1983) *The Problem of Sociology*. London: Hutchinson.

Luff, P., Hindmarsh, J. and Heath, C. (2000) *Workplace Studies: Recovering Work Practice and Informing System Design*. Cambridge: Cambridge University Press.

Maynard, D. (1991) Interaction and asymmetry in clinical discourse, *American Journal of Sociology*, 97, 2, 448–95.

Mechanic, D. (1961) Sources of power of lower participants in complex organisations, *Administrative Science Quarterly*, 7, 349–64

Melia, K. (1987) *Learning and Working: the Occupational Socialization of Nurses.* London: Tavistock.

Mizrahi, T. (1986) *Getting Rid of Patients: Contradictions in the Socialization of Physicians.* New Brunswick, New Jersey: Rutgers University Press.

Mumford, E. (1970) *Interns: from Students to Physicians.* Cambridge, Massachusettes: Harvard University Press.

Olesen, V.L. and Whittaker, E.W. (1968) *The Silent Dialogue: a Study In The Social Psychology of Professional Socialization.* San Francisco: Jossey-Bass Inc.

Parsons, T. (1951) *The Social System.* New York: Free Press.

Pilnick, A. (1998) 'Why didn't you just say that?' Dealing with issues of asymmetry, knowledge and competence in the pharmacist/client encounter, *Sociology of Health and Illness*, 20, 1, 29–51.

Powell, W.W. and DiMaggio, P.J. (eds) (1991) *The New Institutionalism in Organizational Analysis.* Chicago: University of Chicago Press.

Samson, C. (2005) The fracturing of medical dominance in British psychiatry? In Annandale, E., Elston. M.A. and Prior, L. (eds) *Medical Work, Medical Knowledge and Health Care: a Sociology of Health and Illness Reader.* Oxford: Blackwell.

Scheff, T.J. (1961) Control over policy by attendants in a mental hospital. *Journal of Health and Human Behaviour*, 2, 93–105.

Sharrock,W. (1979) Portraying the professional relationship. In Anderson, D.C. (ed.) *Health Education in Practice.* London: Croom Helm.

Simmel, G. (1950) Superordination and subordination. In Wolfe, K.H. (ed.) *The Sociology of Georg Simmel.* New York: Free Press.

Stacey, M. (1981) The division of labour revisited or overcoming the two Adams, In Abrams, P., Finch, J. and Rock, P. (eds) *Practice and Progress: British Sociology 1950–1980.* London: George Allen and Unwin.

Stein, L. (1967) The doctor-nurse game, *Archives of General Psychiatry*, 16, 699–703.

Strauss, A., Schatzman, L., Bucher, R., Ehrlich, D. and Sabshin, M. (1964) *Psychiatric Ideologies and Institutions.* London: The Free Press of Glencoe Collier-Macmillan.

Strauss, A., Fagerhaugh, S., Suczet, B. and Wiener, C. (1985) *The Social Organization of Medical Work.* Chicago: University of Chicago Press.

Timmermans, S. (1998) Resuscitation technology in the emergency department: towards a dignified death, *Sociology of Health and Illness*, 20, 2, 144–67.

Timmermans, S. and Berg, M. (2003) The practice of medical technology, *Sociology of Health and Illness*, 25 (Silver Anniversary Issue), 97–114.

Turner, J.H. (1984) Durkheim and Spencer's principles of social organization, *Sociological Perspectives*, 27, 1, 21–33.

Whalen, J. (1995) A technology of order production: computer aided dispatch in public safety communication. In ten Have, P. and Psathas, G. (eds) *Situated Order: Studies in the Social Organisation of Talk and Embodied Activities.* Washington: University Press of America.

Whalen, M. and Zimmerman, D.H. (1987) Sequential and institutional contexts in calls for help, *Social Psychology Quarterly*, 50, 172–85.

Chapter 2

The promotion of private health insurance and its implications for the social organisation of healthcare: a case study of private sector obstetric practice in Chile

Susan F. Murray and Mary Ann Elston

Introduction

In Griffith's (2003) review of studies of the social organisation of healthcare for *Sociology of Health and Illness*, she drew attention to the scarcity, within medical sociology, of empirical studies exploring the links between different levels of healthcare organisation, from policy making to service use. She argued that such studies can illuminate how government reforms become enacted, and at times 'subverted and reshaped', within a given social and cultural context. In this chapter we use a case study of obstetric care in Chile to explore such linkages. We examine the implications of the process of privatisation of a national healthcare system for the delivery, organisation and, ultimately, the outcome of services. We outline the national healthcare system reforms set in train by the military dictatorship in Chile from 1973, and the associated development of new meso-level organisations – private companies in charge of financing and delivering medical services to their subscribers – and their growing importance in Chilean healthcare over the latter half of the 1980s and the 1990s. Then, drawing on obstetricians' accounts, we analyse the work patterns that have been occasioned by changes in healthcare financing. Finally, we consider some consequences of the changes for service delivery and for users of private maternity care.

Macro-level changes in healthcare policy in Chile
Chile was one of the first countries to provide comprehensive medical coverage for some categories of non-military public workers from 1918, and the Chilean National Health Service was created in 1952. From 1952 until 1973, Chile moved progressively towards increasing publicly-financed and mainly publicly-provided healthcare. By the early 1970s, the National Health System and SERMENA (a government administered health insurance plan for white collar workers and their dependents) covered approximately 85 per cent of the population, the rest being mainly covered by employer-based private insurance schemes (Quesney 1996, Scarpaci 1987). Only a very small élite paid directly for private care.

Plans to consolidate all public and private medical programmes into a single system were cut short by the military coup of 11 September 1973. The agenda of the new regime was very different from that of the socialist government it had overthrown. Within two months of the coup, Air Force Colonel and Minister of Health Alberto Spoerer announced that 'healthcare is not given, rather, it must be obtained by the people' (Scarpaci 1987), effectively announcing a strategy of *decollectivisation* (Mohan 1991), a retreat by the state from acceptance of responsibility for healthcare financing and service provision.

Faced with high inflation and economic stagnation in Chile, and influenced by American neo-classicial economists, the Pinochet military government set out explicitly to shrink the public sector (Miranda *et al.* 1995). Government spending on healthcare fell by 40 per cent between 1973 and 1987 (Jiménez de la Jara and Heyermann 1993). During the Pinochet years, increasing private sector participation was seen as a key component in creating a competitive medical marketplace that would eventually reduce the government's role (Scarpaci 1987). From the beginning of the 1980s, the state-controlled healthcare system was gradually replaced by a market-economy model in which the state was allocated a subsidiary role in financing and a much reduced role in providing healthcare (Quesney 1996).

Meso-level change

Thus, while the National Health Service System (SNSS) still provides health services throughout Chile, it was decentralised in 1980 to 26 autonomous health service areas responsible for the operation of all public hospitals and oversight of the primary care system. A new financial institution, FONASA (the National Health Fund), was created to manage the public sector resources generated through compulsory deductions from workers' pay, central government contributions and income from user charges.

The other key change in the administration of healthcare was the creation of the ISAPRE (Instituciones de Salud Previsional – Institutions of Health Provision) in 1981. New legislation enabled salaried workers to opt out of contributing to FONASA and to place their mandatory wage withholdings (4% at that time, 7% by the 1990s) into a private medical programme of their choice. Modelled on US health maintenance organisations, ISAPRE are private companies responsible for delivering medical services to affiliates, 'superseding the State in delivering healthcare benefits and paying common illness subsidies' (Asociación Gremial de Instituciones de Salud Previsional Chile 2003). ISAPRE health plans generally offer affiliates a choice of medical services through a network of providers who have agreements with them. ISAPRE are funded through affiliates' payroll deductions, used to pay monthly premiums that vary according to the specific coverage plan chosen. There are further co-payments at the time of service use, typically from 10– 30 per cent of costs.

The National Association of ISAPRE describes the ISAPRE system as one that 'allows private individuals to organise and take charge of managing a healthcare system' (Asociación Gremial de Instituciones de Salud Previsional

Chile 2003: 6). The ISAPRE healthcare programmes were intended to embody the spirit of a 'free-market society' because consumers could select pre-paid medical plans based upon their preferences, demographic traits and medical needs (Miranda *et al.* 1995). ISAPRE do not charge standard fees. Each one specifies different charges for the medical services covered, and plans differ in coverage, providers and price.

Initially, in the early 1980s, enrolments in ISAPRE were more sluggish than expected. Plans were relatively expensive, and provisions for contributors taking maternity and medical leave from employment were highly restrictive, but a subsequent series of government decrees supported ISAPRE growth. In 1986, new legislation made enrolment for women without paid employment less expensive. The government also created a special fund (Fondo Unico de Prestaciones Familiares) to handle subsidies for maternity leave (Miranda *et al.* 1995). Then, in late 1986, the state began contributing an extra two per cent towards the mandatory wage withholding (now 7%) for all low-income workers. This subsidy (removed again in December 1999) encouraged even some relatively low-income workers to join the ISAPRE by choice rather than rely on state healthcare provision (Miranda *et al.* 1995).

As a result, the ISAPRE private health insurance system began to grow rapidly in the second half of the 1980s. In 1982 less than two per cent of the population was affiliated to an ISAPRE (with 85% of the population contributing to FONASA). By 1986, ISAPRE coverage had risen to over seven per cent of the population (FONASA 79%), and by 1993 it had reached 25 per cent (FONASA 65%)(Ministerio de Salud 1996). Initially, ISAPRE offices and subscribers were geographically concentrated in the upper-income neighbourhoods of Santiago and other principal cities. But, later, encouraged by the increased government subsidy, many ISAPRE began to recruit more widely among the urban population (Miranda *et al.* 1995). By the late 1990s, ISAPRE membership extended to 27 per cent of the population. The absolute number of 'beneficiaries' of the ISAPRE peaked in 1997. At this point, in a very different political climate, the Ministry of Health began encouraging low-income subscribers to return to the state system, removing the state subsidy for ISAPRE contributions (*Latin America Weekly Report* 1998), and by 2003, the proportion of the Chilean population covered by the ISAPRE system had fallen to 20 per cent (Asociación Gremial de Instituciones de Salud Previsional Chile 2003). But, as ISAPRE have favoured those least likely to need expensive medical care, the young and healthy are heavily over-represented among those covered. Some 70 per cent of ISAPRE subscribers were aged under 40 years in the late 1990s (Bertranou 1999: 24).

The second part of this chapter will explore some of the implications that this new system had for the organisation of obstetricians' work. Scarpaci (1987) suggests that the organised medical profession was not itself a major force promoting the privatisation of medical services in Chile. The Ministry of Health, mainly headed by members of the armed forces during the Pinochet years, largely bypassed the Colegio Médico (the Chilean medical society) in its policy formulation. Ideologically, the military government was opposed to producer dominance and professional monopoly. Nonetheless, in practice, the growth of the private insurance sector benefited the medical profession relative to other healthcare occupations. In their health plans and reimbursement rules the ISAPRE operated policies of exclusionary closure (Witz 1992). In the case of maternity care, obstetricians, not midwives, were established as lead carers for ISAPRE reimbursement purposes.

Maternity care in Chile
Childbirth is more highly institutionalised in Chile than in much of the rest of Latin America, home birth being rare. By 1992 a professional attendant (usually a midwife or doctor) was present at the birth of 99.2 per cent (276,987) of babies registered as born in Chile (Ministerio de Salud/Instituto Nacional de Estadisticas Servicio de Registro Civil e Identificación 1994). Obstetrics and Gynaecology is a recognised post-graduate medical speciality, although doctors without this qualification are not barred from attending deliveries. Chile is one of the few Latin American countries which has retained midwifery as a distinct occupational grouping – the *matronas* (Szmoisz and Vartabedian 1992, Faúndes and Cecatti 1993). Chile also had, in the 1990s, the distinction of having the highest caesarean section rate in Latin America and possibly the world (Bélizan *et al.* 1999, Walker *et al.* 2002).

Private health plan coverage of maternity care
Given the emphasis on attracting the young and healthy, the sector of the Chilean population most likely to want to access maternity care is heavily over-represented among ISAPRE subscribers and their dependents. In 1994 almost half – 49 per cent – of the total 'beneficiaries' within the ISAPRE system were women. Women comprised 31 per cent of direct subscribers (*cotizantes*) to the ISAPRE, and 62 per cent of the dependents covered (*cargas*) according to Ramírez and López (1995). ISAPRE are obliged by law to provide antenatal care, and postnatal care up to six months, within their packages of care. This legal obligation does not extend to delivery care and cover for the latter is often subject to restrictions or exclusion clauses. Nevertheless, by the end of the 1990s, nearly one-third (31%) of obstetric deliveries in Chile were covered by the ISAPRE sector (Ministry of Health figures for 1998, in SERNAM 2001).

An important distinction, with respect to maternity care provision, between women covered by an ISAPRE plan and those who rely wholly on the public health service is that the private healthcare plan provides for continuity of care from a single, named personal obstetrician. All private patient births, whether or not covered by ISAPRE, are attended by the personal obstetrician with his/her team. Private antenatal and postnatal care is provided at obstetricians' own consulting rooms. These are often located, as are delivery suites, at one of the many private maternity facilities which have mushroomed in Chilean cities with the growth in private insurance. Public and teaching hospitals also have a range of facilities for private patients. In contrast, women relying on the national health service system receive antenatal and postnatal care from duty midwives and doctors at health centres on a first-come first-served basis. They then attend their local public hospital for labour and delivery care. In the public labour wards of national health service hospitals, vaginal deliveries are attended by midwives, and doctors deal with complications.

The rest of this chapter draws on obstetricians' accounts to explore key features of the social organisation of healthcare work that emerged within the substantial new privatised sector by the 1990s.

Methods

The data for the following analysis are drawn primarily from interviews with 22 obstetricians, which were conducted in Santiago between 1995 and 1997, when affiliation to the ISAPRE system was at its peak. They were carried out by SFM, a Spanish-speaking British researcher with familial connections in Chile. These interviews explored how private maternity care was viewed, patterns of engagement with private maternity care, problems encountered by the respondents within the private care system and strategies to deal with these, and attitudes to pregnancy, childbirth and mode of delivery. The interviews, lasting 40 minutes to two hours, were audio-taped and transcribed verbatim in Spanish. All transcriptions were entered into a qualitative software programme (Nvivo) for coding and analysis. Constructs and concepts were identified from each interview by the same researcher. Similarities and differences were examined, and underlying themes and categories identified with which to analyse the data. Interpretation was also informed by parallel sets of interviews conducted with midwives, and with women clients. The findings were written in English.

The aim of the sampling approach was to include a wide range of age, experience and professional opinion among obstetricians. All interviews were carried out in Santiago, where one-third of the country's population and almost two-thirds of its doctors live (Instituto Nacional de Estadísticas 2000). Ministry of Health approval for the study was obtained. However, the principal mechanism for recruitment of interviewees was personal introduction, in keeping with Chilean social expectations. Seven obstetricians were recruited from a university teaching hospital, via personal introductions by an interested colleague. They ranged in experience from a newly qualified obstetrician to the head of a perinatal medicine unit, and had varying levels of involvement in private sector work. One senior doctor declined to be interviewed. A further seven obstetricians were then recruited from the public hospital serving a poor sector of the city. Here, a presentation about the study made by SFM served as a means of introduction to future interviewees, who were then selected to represent doctors at different stages in their career.

This group included two trainee obstetricians, and the one senior doctor who had no private patients of his own.

To widen the sample, a further eight interviewees were then approached through personal contacts – six via doctors previously interviewed, and two via personal introduction by friends who had been their patients. They included the author of a widely used obstetrics textbook, and a senior obstetrician at another university hospital in Santiago. Four interviewees were recruited who worked solely within the private healthcare sector, three as staff members at exclusive private clinics, the fourth being a 'freelance' obstetrician/gynaecologist, who had no institutional base, using the facilities in six or seven clinics as required. Finally, an obstetrician who attended home births and 'water births' in his private sector work agreed to be interviewed.

Thus interviewees included young obstetricians for whom the current organisation of work was all that they had experienced, as well as those who had experienced the changes at first hand. Some key characteristics of the obstetrician respondents are summarised in Table 1. The predominance of male obstetricians in the sample reflects the composition of this specialty in Chile.

Table 1 *Characteristics of respondents of tape-recorded interviews – obstetricians sample (total 22)*

Sex	Female	1
	Male	21
Years of experience in speciality	In training	2
	Newly qualified	2
	3–10 years	6
	11–25 years	10
	> 25 years	2
Sectors of work	Public (maternity hospital /or health centres) plus private sector work	9
	University hospital plus private Sector work	9
	Private sector only	4
Average number of private births per month	Births	
	0	1
	1–2	4
	3–5	6
	6–8	2
	9–11	5
	12–14	4
Family commitments	Single	4
	Married, no children or grown up	2
	Separated, children living elsewhere	2
	Family with children	14

Findings

Dual practice
Most (18 of 22) of the obstetricians interviewed were engaged in 'dual practice': that is, they had salaried 'base posts' in the government and/or university healthcare sector and also did private practice. The structure of the Chilean obstetricians' time commitment in their base posts (often mornings only, plus one 24-hour shift per week) easily permits private practice, and many clinicians alternated between their public and private sector work in the course of a normal working day. The doctors' private practice consulting rooms may be within a health facility or in separate office-style accommodation, with private laboratory and ultrasound facilities close at hand. The location and décor of doctors' rooms reflect the social status of the client group for whom the obstetrician caters. There is no Santiago equivalent of London's iconic Harley Street as the centre of élite private healthcare consulting rooms (Humphrey 2004). For an aspiring obstetrician, however, an élite address on the right side of town is an important factor in generating a desirable *clientela* (private client base).

The primary reason obstetricians gave for doing private maternity care work was financial remuneration (highlighted in all 22 accounts), and there was no reticence in admitting this – unlike that found among UK dentists interviewed about their balance between public and private work (Calnan *et al.* 2000):

> I do private work because the salaries in the hospital are so low . . . and one can earn a thousand dollars in one (private) birth (Obstetrician 03).

Interviewees working in the public or university sectors saw private sector work as a necessary and inevitable element of their personal income generation. Privatisation of the Chilean public sector had also affected education, and private practice was seen as an economic necessity for those doctors with families and with school and university fees to pay. The type of private work undertaken varied among individuals, but even those doctors still in training in their speciality had their own private practices. One senior doctor no longer took private maternity clients of his own since he had been involved in a difficult litigation case at the public hospital. However, he still assisted colleagues on occasions as 'second surgeon' in operations within the private sector.

Apart from its financial benefits, the single named obstetrician model appealed to the doctors, as it drew on classic features of professional autonomy: solo practice, highly individualised clientele, little external control over clinical decision-making (Elston 1991). In-depth interviews with women clients, discussed elsewhere show, furthermore, that this model is also prized by subscribers, for at least two reasons: firstly, personalised care denotes a high status for the recipient of private obstetric care, being recently only the privilege of the wealthy; and, secondly, it is perceived as bringing better guarantees of individualised and high-quality care (Murray 2002).

As in the US, where obstetricians 'may suffer the time and energy demands of delivering babies in order to feed their gynecologic practice' (Keeler and Brodie 1993: 368, see also Baumgardener and Marder 1991), the Chilean obstetricians felt some pressure to accumulate private maternity patients. This was not solely financial. As many private clients arrived by personal recommendation, the number of *clientela* a doctor had was a visible measure of popularity and professional success, much like a UK surgeon's waiting list (Pope 1991). For those aspiring to join the handful of exclusive private clinics as staff members (see below), proof of their ability to generate a substantial *clientela* was an important bargaining tool.

Some obstetricians were well aware of the dangers of accumulating too much private work:

> Medicine is a very absorbing profession, above all the private aspect. I'd say the private aspect is more enslaving than the hospital. In the hospital one has set hours of work. In the private area it isn't like that, and once you enter the machine it is impossible to get out. Your economic resources increase, your needs increase, and then to meet your needs you have to carry on earning more money. You can never leave off, and you lose the other things in life (Obstetrician 12).

One interviewee said that he valued the way in which the structure of private work allowed him to give continuity of care; another that it permitted long-term relationships with patients over many years. But this point was not mentioned spontaneously by any other interviewees and there was little emphasis from obstetricians on any intrinsic rewards from developing personalised relationships with private sector clients.

Obstetricians' relationships with the insurance funds

Attitudes to the ISAPRE system of private health insurance ranged from critical, through ambivalent, to accepting. The large-scale entry of private insurance funds into the healthcare market may have opened up the scale of private work available to obstetricians. But, for some, it has also brought new constraints on their private practice organisation. All the ISAPRE operate networks or rolls of 'approved' or 'preferred' healthcare providers (*Red de Prestadores Preferentes*) with whom they have agreements to provide services to the subscribers of their Health Plans. This facilitates the voucher (*bono*) system for co-payments, in which the subscriber pays the co-payment at the rate set by the ISAPRE in exchange for a voucher, which she presents to the doctor at the time of service. The doctor then claims the payment from the ISAPRE. This arrangement can provide the doctor with a certain amount of custom, but it binds the doctor to charge the rates set by the fund concerned.

To cover the market fully, most ISAPRE also offer a range of plans in which the client is also permitted to use care providers and facilities outside the 'preferred' network – for example, if a woman had established a relationship with a certain obstetrician before joining that plan, or wished to use a facility or service near her home. In these cases the patient pays the whole amount 'up front' and claims back the amount that is covered by the ISAPRE health plan. In this situation doctors have more flexibility in their charges, the constraint being what clients are able and willing to pay.

Where obstetricians position themselves in this insurance coverage market place depends upon their reputation, experience and contacts. Most expressed frustration at being constrained by the reimbursement rates set by the insurance funds. Only a minority of 'jet-set doctors' (the term used by several interviewees), with the right social as well as clinical credentials, could charge whatever they wanted, knowing that their clients would pay. Being able to charge 'one's own rate' was therefore a symbol of a special status among obstetricians.

Work 'signatures'
Because of the way in which the specialty is defined within Chilean medicine, interviewees' private sector work, like their public sector work, usually spanned both obstetrics and gynaecology. The balance of their commitment to one or the other, and to sub-specialties within them, such as gynaecological oncology, family planning or adolescent reproductive health, varied from doctor to doctor. Doctors had their own individual profiles or 'signatures' (Wennberg *et al.*1982, Pope 1991) for their private sector case mix, as well as for their public/private sector commitments, as discussed. Within private practice, maternity cases (*i.e.* from first antenatal visit to postnatal care) were seen as time-consuming, but a good source of income. The average number of private births the obstetricians interviewed said that they currently attended varied substantially – from none (the obstetrician who had been involved in litigation) to 14 per month (see Table 1). Thus, obstetric and gynaecology work in the private sector takes different forms, as illustrated in the examples in Table 2. But the end result for an individual obstetrician is often a fairly complex and fragmented work schedule, providing different types of care under different payment schemes and over several locations, in which work has to be organised as systematically as possible.

The social and occupational hierarchy within obstetrics
A social and occupational hierarchy within the private sector was in evidence in terms of the rewards accruing to obstetricians. Those mainly working in public sector hospitals, who had a few private clients in order to generate supplementary income for themselves, had a quite different experience of the private sector from those who were 'staff members' at expensive private hospitals. Many of the interviewees expressed some degree of 'sentimental attachment to the concept of a collective public service' similar to that of specialists engaged in dual public and private sector practice in England

Table 2 *Examples of work schedules of obstetricians across different sectors of maternity care in Santiago*

Obstetrician 14
Staff member at public hospital
Four mornings a week at public hospital, plus weekly 24-hour shift
4 weekday afternoons at private medical centre doing consultations and ultrasounds
18.30 to 20.00 hours, runs own private clinic
Booked operations between 20.00 hours and 23.00 hours
Averages 10 private births per month

Obstetrician 16
Staff member at public hospital
4 mornings a week in public hospital in the city
24-hour shift in another (rural) public hospital
Own private clinic 3 afternoons from 15.00 to 20.30 hours at consulting rooms
Averages 2 private births per month

Obstetrician 20
Staff member at exclusive private hospital
4 half-day shifts doing consultations and 4 doing ultrasound, at this hospital
On call system for nights one week in 16 for emergencies
Attends 'own' births, conducts surgery, research, at same facility
Two update meetings per week. With other staff
Averages 10 private births per month

Obstetrician 02
Staff member at university hospital
Weekday mornings – Clinical teaching and administration at several sites within university teaching hospital
Weekly 24-hour shift at armed forces hospital
Private consultations 3 afternoons/evenings per week
Averages 2–4 private births per month

Obstetrician 05
Freelance private work only
Operates 3 mornings per week
Private consultations from 11.00 to 13.00 and then from 16.00 to 20.00 hours then visits surgery patients in the clinics/hospitals (has 6–7 clinics/hospitals where he does work)
Averages 5 private births per month

Source: Tape-recorded interviews with obstetricians

(Humphrey and Russell 2004), and they still spoke of their private work as the *extra sistema* (outer-system) or simply as *afuera* (outside). For some, however, a clear career progression was apparent (or desired if not yet achieved) away from public or university sector work towards that of the private-for-profit sector. Having gained sufficient experience, a few obstetricians leave their public sector jobs and work completely 'outside'. This may take a number of forms. Some work 'freelance' attending only their private clients and perhaps doing work on a sessional basis in a private medical centre. Those who have been highly successful, and are sufficiently well connected socially, may be invited to join one of the more exclusive, well-resourced private clinics as a staff member. This option, seen as the pinnacle of professional and social achievement by many of the interviewees, is only open to, quite literally, the chosen few (the 'jet set'), as explained in the following interview extract:

> The Clínica 'B', in terms of its medical team, is completely closed. It is a closed clinic and entry to this clinic doesn't generally start from some doctor wanting to join the clinic, but rather from the group of doctors who are here, in this case the obstetrician and gynaecology group, deciding to incorporate some more doctors. They look around among the possible candidates; they investigate which doctor seems to them to combine the characteristics required here. In general they look for doctors who have more or less finished their specialist training and who have shown themselves in a preferential position in the national circles of the medical community. Apart from all the other factors that have to be present – education, good connections . . . And usually they take doctors who have been closely connected in some way to the (Catholic) University (Obstetrician 20).

Being 'on the staff' of such a clinic does not imply a salaried post. The Latin-American clinic model typically has a contract physician panel that 'assumes the financial risk' (Stocker *et al.* 1999). In other words, these staff obstetricians still have to generate their own income, through their personal private clients who have private healthcare plans. Aspiring obstetricians would court such clinics in the hope of acceptance, but the strength of social oligarchy and the *cupula*, the incumbent holders of power and authority, made it hard for outsiders – however ambitious and talented – to penetrate. One doctor interviewed was on the verge of despair, and talked about giving up medicine. Despite having built up a good-sized *clientela* and establishing himself as a 'bright young doctor' with research activities at a university hospital, he felt the doors to the élite clinics were closed to him because he belonged to a minority religious group.

Dealing with market vulnerability

Most of the interviewees recognised that they did not belong to the social world of the closed clinic, and their decisions on the relative size of their public or private sector commitments were pragmatic ones. Leaving behind a public or university sector post was considered risky by most interviewees. Wages in the public sector might be low, but provided a degree of economic security, unlike the private sector which:

from the economic point of view it is just like a business, a shop . . . the day that you don't hold a surgery (*consulta*), you simply don't earn any money, that's the worry. . . . (Obstetrician 19).

There is a plentiful supply of obstetricians in the capital city, and many of them felt vulnerable to the market nature of their private sector activities. They spoke about competitive pressures and about the need to keep clients happy lest they went elsewhere. Their accounts displayed an ambivalence towards their situation, as did the discrepancies between their own practices on the one hand and what they deemed an acceptable attitude from the client on the other. The use of market mechanisms to keep the volume of private clientele down to a manageable size – for example, by raising personal fees to a level that excludes some of the potential clients – was reported with pride by some doctors. But they were disapproving of clients who acted, or who expected to act, as consumers. Distaste and offence were expressed in situations where clients alluded to the monetary nature of the relationship. Two obstetricians made comments about the husbands of clients expecting to 'be able to buy a healthy baby by the kilo'. Some felt that they themselves had been commodified by the private system, complaining that women now went 'window-shopping' (*vitrineando*) [Obstetrician 01], trying out different doctors 'like you look for shoes in the shoe shops' [Obstetrician 22]. One way of dealing with this discomfort was to attempt to assert control by redefining the parameters, subjecting women's option to change provider to specified conditions that they themselves dictated:

The doctor-patient relationship must be optimal, I always say to my patients that if one day they lose confidence in me, I prefer that they change doctors immediately (Obstetrician 02).

Another tactic to pre-empt this was to insist on obedience from the patient:

If I give instructions and the patient doesn't follow them, I cannot assume responsibility for her pregnancy and birth. If she doesn't go to the check-ups when she needs to go, if she puts on five kilos in weight at every check up, if I send her to a specialist for evaluation of a problem and she doesn't go, I can't go on. In such a case there isn't a good response from the patient, and I could not have confidence in that patient, as a patient (Obstetrician 08).

Despite such heroic statements (which may not reflect actual behaviour with clients), the obstetricians fully recognised women's expectations of personalised care from their *medico de confianza* [literally 'trusted doctor'], and how integral this was to the success of the private care relationship:

One must make the patient feel, once she is coming to you for antenatal care, that one is permanently at her disposal, for whatever emergency, whatever the hour, whatever the day (Obstetrician 01).

The Chilean obstetricians recognised that an explicit part of the 'sentimental' work (Strauss *et al.* 1982, 1985) in private obstetric care was their guarantee of direct personal availability at all times (as found in Tanassi's (2004) study of childbirth in Italy). Accordingly, they furnished their clients with the numbers of their personal bleeper and office, home and mobile phones. It was also felt that this personalised relationship with the client was the obstetrician's best protection against the bringing of a malpractice suit if something was to go wrong. This echoes the professional literature in the US and the UK reviewed by Annandale (1989) in her study of the 'medical malpractice crisis'. (Litigation cases are infrequent in Chile, although when they do occur they receive widespread media coverage.)

Managing 'whatever the hour, whatever the day' (and wherever)
The investment in and diversification of the private sector since the 1980s meant that one doctor might hold clinics, attend births, perform gynaecological operations, etc. in a number of different, often scattered, locations on the same day. Most private clinics are 'open' to individual obstetrician gynae-cologists for their patients' labour care or for operations. Such clinics vary in their standard of accommodation, equipment and, consequently, in cost and the proportion of a care episode covered by a given ISAPRE plan. Although obstetricians may state their preferences, it is the client who chooses where to have her delivery or operation. So all but the most favoured and fashionable obstetricians, who have achieved staff status at an exclusive clinic, have to be prepared to travel from one facility to another, through Santiago's often very congested traffic.

These work schedules are demanding and leave little space for personal and family life. Many of the doctors interviewed indicated that they would have preferred not to have to 'run around Santiago like a lunatic' and would have much preferred to work only in one institution. But this option was open to few. Most described themselves as trapped by the standard of living offered by private obstetric practice:

. . . when you talk to the oldest doctors, whom you think must be coming out of this running between births, they carry on getting up at five in the morning. I think it is everybody's dream, at the beginning, to say I am going to do this for a few years only, then no more. You dream like that when you start out, but you have to carry on until you retire (Obstetrician 01, university hospital and private work).

The commitment of availability 'whatever the hour and whatever the day' (Obstetrician 01) to various clients was an organisational challenge, given fixed commitments such as consulting hours, hospital rounds and sessional bookings. Trips further afield than a few hours from Santiago could be difficult. Most doctors felt that they had to seek some sort of equilibrium between the demands of their private work, the requirements of their 'base job' and their personal lives.

Obstetricians resolved or reduced the stress from conflicting demands by one or more of three strategies (Murray 2000). They could limit the number of private clients by increasing fees. They could attempt to concentrate their work in fewer locations, ideally joining the staff of one of the exclusive private hospitals and working only there. Finally, they could programme the births of their private clients: either by induction and/or regulation of labour with oxytocins, or by elective caesarean section as illustrated in the following interview extracts:

One tries to place the delivery care in this timetable in the evening, that is to say what is called a programmed birth. . . . (Obstetrician 14).

It's a good thing for the doctor, in order to organise himself. As these are private births, and the great majority of colleagues have their institutional work as well, they can't be going off everyday, running out at dawn, or running out from the hospital (Obstetrician 16).

To be with her, I programme a day when I have plenty of time – to be able to concern myself with her labour . . . It has to be a day when I don't have clinics (Obstetrician 17).

Programming of the commencement of labour in the private sector is an attempt to regulate the timing of the birth of the baby – an event at which obstetricians are committed, by virtue of the personalised doctor-patient relationship, to assist in person. Data from a postnatal survey conducted in a public hospital, a university hospital and a middle-income private clinic (Murray 2002), confirm that a woman's chances of having a non-programmed birth can vary substantially according to the sector in which she has her delivery care. Two-thirds of women receiving care from the public sector started in labour spontaneously and also delivered their babies without recourse to caesarean section or to forceps. In all of the groups in that survey who received private obstetric care, however, only a small minority of women had this experience.

Programming birth by induction of labour, however, can be a long drawn-out and unpredictable process. It is not always successful and the course of labour may not always dovetail with other work commitments. The alternative of elective caesarean sections has the appeal of greater reliability and a shorter time commitment:

> If I know that with a caesarean it is going to take me two hours at such and such a time . . . I programme it into my timetable and I don't lose other things (Obstetrician 06).

Thus, elective caesarean sections were seen as a more reliable and attractive option for the many obstetricians interviewed who did not have strong objections to non-medically indicated caesarean sections. These could be easily scheduled for a time that was convenient for the various members of the private childbirth care team (obstetrician, paediatrician, anaesthetist, midwife), for the facility and for the clients. They allowed doctors to rationalise their activity, even allowing some, according to one interviewee, to perform three caesareans in a morning. A small number of interviewees considered such behaviour professionally unethical (doctors allowing self-interest to influence their medical decisions), or morally unacceptable (such caesareans were 'unnatural', and led to artificial limitation of family size). This latter position, taken by those doctors who might be described as 'activist Catholics', was the most strongly expressed. But these *vaginalistas* as they referred to themselves, were very much in the minority among interviewees. The *operadores* or *cesareanistas* expressed their belief that caesarean birth under epidural was a safe and relatively simple procedure, and therefore unproblematic. These doctors saw themselves as 'modern' rather than traditionalist, and were more likely to be willing to engage in a discourse that accepted social reasons for clinical actions. They saw programming as advantageous not only to themselves, but also to their clients, because it helped them to avoid the surcharges levied for services provided outside normal working hours:

> [we] programme this on an appropriate date, even though it brings it forward slightly, so as not to have a delivery in the early hours of the morning, or on a public holiday – for economic reasons . . . For the private patients . . . a delivery in the night or on a Sunday comes out twice as expensive (Obstetrician 16).

However, although concerns for clients' circumstances were expressed in such terms, the explicit discourses of women's choice in childbirth practices that are found elsewhere in the debates around elective caesareans (Paterson-Brown 1998, Showalter and Griffith 1993, Wax *et al.* 2004) were largely absent from these Chilean obstetricians' accounts.

The work contexts of individual obstetricians made them more or less vulnerable to the pressures to operate. A statistical comparison *within* a well-known and fashionable private clinic in Santiago revealed far lower caesarean section rates (28%) for the clients of obstetricians who were 'on staff' at the clinic, and who had presumably therefore managed to concentrate their *clientela* into one place, than for the clients of visiting doctors, who had not (57%) (Schnapp and Sepulveda 1997).

Discussion

Funding mechanisms can have a powerful influence on the organisation of healthcare work and ultimately on the management and outcomes of care. But increased privatisation of health service funding does not in itself dictate the particular organisational form through which privatisation is enacted or the ensuing forms of service provision. According to these Chilean obstetricians' accounts, the dynamics somewhat inadvertently created through the shrinkage of the public sector and the promotion of new organisational structures for healthcare financing propelled individual obstetricians into a new and demanding engagement with private practice. In a context in which the organised character and political influence of the medical profession had been greatly diminished during the period of military dictatorship, there was no scope for collective professional resistance to the fragmentation of work created by diversification in the healthcare market. In a situation of deteriorating public sector salaries and relative oversupply of obstetricians in Santiago, most obstetricians sought private maternity clients as this is lucrative work.

For a significant proportion of Chilean women seeking maternity care, privatisation has led to expanded access to ostensibly highly personalised relationships with specialists, *i.e.* to what Strong (1979: 207) termed the high level of 'product differentiation' characteristic of private format doctor-patient relationships. The ISAPRE health plan system, however, mediates between obstetricians and their clients in complex ways. The client has a considerable degree of choice between and within plans, for example, with respect to obstetrician and place of delivery, and therefore, over the cost of private professional care for a pregnancy. But thereafter, both ISAPRE and clients generally cede decision-making to obstetricians. The discourses of women's choice in childbirth practices, or of natural childbirth as an ideal, appear to have little legitimacy in the mainstream of private obstetric care in Chile.

For most obstetricians, conflicts between the obligation to provide individualised care for their private clients and to 'be there for them' in person when appropriate, and the demands of fragmented work schedules over several locations present themselves with particular force, given the unpredictability of onset and duration of unmanaged labour. The qualitative data demonstrate how the requirement to provide personalised care, including 'delivering' the baby, and the relatively weak market position of most obstetricians has encouraged highly technologised obstetric practices in the private sector. This was reflected in the national statistics on caesarean section rates at the start of the study (Murray and Pradenas 1998, Murray 2000). Women whose care was covered by ISAPRE plans were twice as likely to have a delivery by caesarean section as women covered by the National Insurance Fund (59% v 28.8% births, 1994 figures, Ministerio de Salud 1996).

There is a large and diverse international literature on inter- and intra-national variations in caesarean section rates. Within the Latin American continent,

Brazil has been the main focus of study on this issue. Recent micro-level studies conducted there have shed insights on care provider behaviours in clinical settings and on the complexities of the notion of service users' 'choice'. De Mello e Souza (1994) used an analysis of interview transcripts, popular newspaper articles and medical literature to examine the formal and informal mechanisms used by Brazilian obstetricians to reinforce conformity to an operative birth culture. More recently, the studies of Brazilian women's view by Hopkins (2000) and Potter *et al.* (2001) have challenged notions that the high caesarean birth rates in Brazil are demand driven, and that operative delivery has widespread popularity among Brazilian women. Relatively little attention, however, has been paid to the meso-level of healthcare systems and to policy and organisational processes in these contexts.

Time management – 'convenience incentives' – have been identified in other marketised healthcare systems such as that of the US. Studies of births using regression analysis on large databases, found indications that physicians in the US may perform caesarean sections to manage their time better (Burns *et al.* 1995, Tussing and Wojtowycz 1993). What the qualitative investigation reported here suggests, however, is that specific features of the organisation of work in the Chilean private maternity care system – solo named practitioners explicitly provided for in private insurance plans, obstetricians' fragmented work patterns across geographically disparate settings and competing clinics, their income vulnerability and weak professional organisation – encourage obstetricians' use of technological intervention and impact markedly upon women's birth experiences.

Conclusions

In this chapter, we have attempted to take up Griffiths' (2003) recent challenge to medical sociologists to address the linkages between the macro-, meso- and micro-levels of health service organisation. In focusing on the government-initiated restructuring of Chilean health services that followed the military coup in 1973, we have chosen a case study where the macro-level might accurately be described as an exercise in 'breakthrough' rather than 'technical' politics (Thompson 1986). That is, the reforms presaged 'major changes in the role of the medical sector', rather than marginal adjustments in the interests of improving healthcare (Hunter 1990: 217). Hunter deployed this distinction more than a decade ago when seeking to encourage medical sociologists in Britain to pay more attention to the meso-level of healthcare systems, to the intermediate layer where policy and organisational and managerial processes tend to be concentrated. Hunter suggested that such attention was especially apposite in the context of 'technical politics' of healthcare (1990: 215). Our analysis, however, has shown the value of analysing the meso-level organisational and policy processes, in this case, of the ISAPRE and the associated patterns of obstetric care organisation, in a context of 'breakthrough' politics, and linking these to the micro-level of service delivery.

Through this analysis we have also illuminated the ways in which privatisation has come to be associated with a marked increase in the caesarean section rate in one country. It is not part of our argument to suggest that this is a necessary or universal relationship. Rather, we hope to have shown the value of examining closely what constitutes 'privatisation' in organisational terms in a particular cultural and social context, so that one of its unintended consequence can be explained.

Acknowledgements

Thanks are due to the participants, who gave us their valuable time from busy schedules. This study is part of a project funded by the Department for International Development, UK (formerly ODA), Grant RD352. However, the views expressed are those of the authors and do not represent the views of DFID. We would like to acknowledge the valuable contribution of Sra Fanny Serani of the University of Chile who was research assistant and advisor to the larger project. Susan F. Murray was lecturer at the Institute of Child Health, University College London at the time of the data collection.

References

Annandale, E.C. (1989) The malpractice crisis and the doctor-patient relationship, *Sociology of Health and Illness*, 11, 1, 1–23.

Asociación Gremial de Instituciones de Salud Previsional Chile (2003) ISAPREs, The private health sector in Chile [online] Available: http://www.Isapre.cl/documentos/isapre.en.pdf [Accessed: 15/06/04].

Baumgardener, J.R. and Marder, W.D. (1991) Specialization among obstetrician/gynecologists: another dimension of physician supply, *Medical Care*, 29, 272–82.

Bélizan, J.M., Althabe, F., Barros, F.C. and Alexander, S. (1999) Rates and implications of caesarean sections in Latin America: ecological study, *British Medical Journal*, 319, 1397–400.

Bertranou, F.M. (1999) Are market-orientated health insurance reforms possible in Latin America? The cases of Argentina, Chile and Colombia, *Health Policy*, 47,19–36.

Burns, L.R., Geller, S.E. and Wholey, D.R. (1995) The effect of physician factors on the cesarean section decision, *Medical Care*, 33, 4, 365–82.

Calnan, M., Silvester, S., Manley, G., and Taylor-Gooby, P. (2000) Doing business in the NHS: exploring dentists' decisions to practise in the public and private sectors, *Sociology of Health and Illness*, 22, 6, 742–64.

De Mello e Souza, C. (1994) C-sections as ideal births: the cultural construction of beneficence and patients' rights in Brazil, *Cambridge Quarterly of Healthcare Ethics*, 3, 358–66.

Elston, M.A. (1991) The politics of professional power: medicine in a changing health service. In Gabe, J., Calnan, M. and Bury, M. (eds) *Sociology of the Health Service*. London: Routledge.

Faúndes, A. and Cecatti, J.G. (1993) Which policy for cesarean sections in Brazil? An analysis of trends and consequences, *Health Policy and Planning*, 8, 1, 33–42.

Griffiths, L. (2003) Making connections: studies of the social organisation of health-care, *Sociology of Health and Illness*, 25, Silver Anniversary Issue, 155–71.

Hopkins, K. (2000) Are Brazilian women really choosing to deliver by cesarean? *Social Science and Medicine*, 51, 725–40.

Humphrey, C. (2004) Place, space and reputation: the changing role of Harley Street in English health care, *Social Theory and Health*, 2, 153–69.

Humphrey, C. and Russell, J. (2004) Motivation and values of hospital consultants in south-east England who work in the national health services and do private practice, *Social Science and Medicine*, 59, 1241–50.

Hunter, D. (1990) Organizing and managing health care: a challenge for medical sociology. In Cunningham-Burley, S. and McKeganey, N.P. (eds) *Readings in Medical Sociology*. London: Routledge.

Instituto Nacional de Estadísticas (2000) Compendio Estadístico. Estadísticas de Educación, Medios de Communicación y de Salud. [online] Available: http://www.ine.cl/chile-cifras [Accessed: 04 September 2001].

Jiménez de la Jara, J. and Heyermann, G. (1993) Re-establishing health care in Chile, *British Medical Journal*, 307, 729–30.

Keeler, E.B. and Brodie, M. (1993) Economic incentives in the choice between vaginal delivery and cesarean section, *Milbank Quarterly*, 71, 3, 365–404.

Latin American Weekly Report (1998) No private care for pregnant Chileans (17 February).

Minsterio de Salud (1996) Información estadistica de prestaciones ortogadas a beneficiarios de la ley 18.469 en sus modalidades de Libre elección e Institucional y Sistema ISAPRE,1986–1994. Departamento de Coordinación e Informatica, Ministerio de Salud, Chile.

Ministerio de Salud/Instituto Nacional de Estadisticas Servicio de Registro Civil e Identificación (1994) Boletin de Nacimientos y Defunciones Año 1992. Santiago, Chile.

Miranda, E., Scarpaci, J.L. and Irarrázaval, I. (1995) A decade of HMOs in Chile: market behaviour, consumer choice and the state, *Health and Place*, 1, 51–9.

Mohan, J. (1991) Privatization in the British health service: a challenge to the NHS? In Gabe, J., Calnan, M. and Bury, M. (eds) *Sociology of the Health Service*. London: Routledge.

Murray, S.F. (2000) Relation between private health insurance and high rates of caesarean section in Chile: qualitative and quantitative study, *British Medical Journal*, 321, 1501–5.

Murray, S.F. (2002) Caesarean birth in the private insurance sector in Chile. PhD thesis, University of London.

Murray, S.F. and Serani Pradenas, F. (1998) Cesarean birth trends in Chile, 1986 to 1994, *Birth*, 25, 3, 207–8.

Paterson-Brown, S. (1998) Should doctors perform elective caesarean section on request? Yes, as long as the woman is fully informed, *British Medical Journal*, 317: 462.

Pope, C. (1991) Trouble in store: some thoughts on the management of waiting lists, *Sociology of Health and Illness*, 13, 2, 193–212.

Potter, J.E., Berquó, E., Perpétuo, I.H.O., Fachel, L.O., Hopkins, K., Rovery Souza, M., *et al.* (2001) Unwanted caesarean sections among public and private patients in Brazil: prospective study, *British Medical Journal*, 323, 1155–8.

Quesney, L.F. (1996) El séctor privado en salud. In Giaconi, G. (ed.) *La Salud en el Siglo XX: Cambios Necessarios*. Santiago: Centro de Estudios Públicos.

Ramírez, C.A. and López, F.D. (1995) Analisis del regimen de ISAPRES desde el punto de vista de los requisitos y servicios a las mujeres. Informe Final (Octubre) (unpublished).

Scarpaci, J.L. (1987) HMO promotion and the privatization of health care in Chile, *Journal of Health Politics, Policy and Law*, 12, 3, 551–67.

Schnapp, C. and Sepulveda, W. (1997) Rise in caesarean births in Chile, *Lancet*, 349, 1029.

SERNAM (2001) Mujeres Chile – estadisticas. Gobierno de Chile Servicio Nacional de la Mujer [online] Available: *www.sernam.gov.cl/estadisticas/* [Accessed:24 November 2004].

Showalter, E. and Griffith, A. (1993) Commentary: all women should have a choice *British Medical Journal*, 319, 1401.

Stocker, K. Waitzkin, H. and Iriart, C. (1999) The exportation of managed care to Latin America, *New England Journal of Medicine*, 340, 14, 1131–6.

Strauss, A. Fagerhaugh, S., Suczek, B. and Wiener, C. (1982) Sentimental work in the technologized hospital, *Sociology of Health and Illness*, 4, 3, 254–78.

Strauss, A., Fagerhaugh, S., Suczek, B. and Wiener, C. (1985) *Social Organization of Medical Work*. Chicago and London: University of Chicago Press.

Strong, P. (1979) *The Ceremonial Order of the Clinic*. London: Routledge.

Szmoisz, S. and Vartabedian, R. (1992) Midwives: professionals in their own right, *World Health Forum*, 13, 291–4.

Tanassi, L.M. (2004) Compliance as strategy: the importance of personalised relations in obstetric practice, *Social Science and Medicine*, 59, 2053–69.

Thompson, F.J. (1986) The health policy context. In Hill, C.E. (ed.) *Current Health Policy: Issues and Alternatives. An Applied Social Science Perspective*. Athens, GA: University of Georgia Press.

Tussing, A.D. and Wojtowycz, M.A. (1993) The effect of physician characteristics on clinical behavior: cesarean section in New York State, *Social Science and Medicine*, 37, 10, 1251–60.

Walker, R., Turnbull, D. and Wilkinson, C. (2002) Strategies to address global cesarean section rates: a review of the evidence, *Birth*, 29, 28–39.

Wax, J.R., Cartin A., Pinette M.G. and Blackstone J. (2004) Patient choice cesarean: an evidence-based review, *Obstetrical and Gynecological Survey*, 59, 8, 601–16.

Wennberg, J.E., Barnes, B.A. and Zubkoff, M. (1982) Professional uncertainty and the problem of supplier induced demand, *Social Science and Medicine*, 16: 811–24.

Witz, A. (1992) *Professions and Patriarchy*. London and New York: Routledge.

Chapter 3

Understanding the social organisation of maternity care systems: midwifery as a touchstone

*Cecilia Benoit, Sirpa Wrede, Ivy Bourgeault,
Jane Sandall, Raymond De Vries and
Edwin R. van Teijlingen*

Introduction

Healthcare systems reflect a complex mix of societal norms, cultural values, government regulations, formal institutional policies and informal practices, tensions over professional boundaries, and social actions of patients and their advocates. These cross-cutting influences present a challenge to those who study healthcare arrangements and design policies to make healthcare more effective and efficient. This chapter advances our understanding by examining the provision of maternity care in four high-income countries: the United Kingdom (UK), Finland, the Netherlands and Canada. All have economies based largely on the free market, social-democratic political systems, and impressively low maternal and infant mortality rates. In addition, all devote substantial public resources to healthcare services and include midwives as providers[1]. Yet these same countries exhibit marked variation in the social organisation of maternity care.

It is true that these four countries have notably different welfare regimes: Canada and the UK are liberal welfare states, Finland is a Nordic universalist welfare state in the social democratic tradition, and the Netherlands is an example of a conservative welfare state of Continental Europe (Esping-Andersen 1990). But, as shown below, these macro-level differences do not fully explain the diversity in maternity care systems.

In our analysis we use midwifery, a female-dominated occupation serving an exclusively female clientele, as a *touchstone*[2] for explaining this variation. The social location of midwifery reveals a society's fundamental cultural ideas about women as (1) autonomous (or not) professionals in the maternity division of labour and as (2) legitimate (or not) recipients of midwifery care services across the childbearing period.

A sociological analysis of midwifery helps us to advance thinking about the organisation of maternity care systems beyond what social or health policy perspectives would allow. It also helps us to better understand the operation of jurisdictional claims in maternity care and the way governments, professionals and clients shape maternity care systems. Our analysis uses a 'decentred approach' to comparative research as developed in *Birth by Design* (De Vries *et al.* 2001). The interdisciplinary team for that project included researchers with 'local knowledge' of the social organisation of maternity care in several Western European and North American countries. One of the crucial reasons for assembling such a team was to move the analysis beyond studies by one or two researchers focused on explaining events in one country by contrasting it to a second or sometimes third case example (Benoit and Heitlinger 1998). Our aim was to 'decentre' the study of maternity care from particular national contexts and to move the analysis to a level where any and all contexts were worthy of sustained examination. Having decentred social contexts, researchers also became uncoupled from their disciplines and particular perspectives and came to share a less ethnocentric and more theoretically sophisticated understanding of healthcare regimes.

A comparative theoretical framework for the study of maternity care

Our framework rests on concepts drawn from three areas of study: comparative welfare states, the sociology of professions and contemporary social movements. We also observe how the gendered composition of groups influences the web of inter-relations between the state, professions and birthing women. We draw attention to differences in the macro, meso and micro organisation of heath care systems across time and place, differences that originate in conflict and negotiation among actors in the maternity care domain (Allen 1997).

Welfare state theory
Welfare state theory defines postwar welfare capitalism as a 'commitment of some sort which modifies the play of market forces in an effort to realise greater social equality for its population' (Ruggie 1984: 11). Feminist scholarship has demonstrated that different welfare regimes have divergent consequences for women. The Nordic states are often called 'women-friendly' because their universalistic system provides women direct access to services as citizens and because social policy there is aimed at both gender and class equality. Women's social rights in conservative welfare states are often defined in terms of their place in the family with the assumption of the male as breadwinner. In liberal welfare states, the state is said rarely to act in favour of women's gendered interests (Lewis 1992).

A central debate among theorists is whether or not the post-industrial, globalised era has created an irreversible decline in developed welfare states, effacing once clear differences in public funding, national regulations and formal policies (Coburn 2001). If this is indeed the case, then maternity care systems in our four case examples should likewise have become more alike over time, with more services being delivered in the private sector, reduced public outlay for maternity services and policies less favourable to women. Others argue that developed welfare states, even during this period of global capitalism, have maintained their distinctiveness in several areas, including the organisation of healthcare (Korpi 2003).

Both the convergence and divergence arguments find support in the development of maternity care systems, depending on the organisational level studied. There has been a growing convergence towards the rationalisation of the maternity division of labour across many welfare regimes; some have emphasised midwife care as a low-tech and therefore low-cost measure. In other cases, rationalisation is sought through the centralisation of births in large hospitals. In order to understand the persisting diversity of maternity care systems, it is necessary to pay attention to the nature of the state and its interests in shaping maternity care in particular ways that may be more (or less) friendly to women workers and recipients of care (Davies 2003).

Sociology of the professions
The sociology of professions draws attention to the unequal relationships among occupational groups within the division of labour and the ensuing struggle over license and mandate. Central to our topic is the power of the medical profession to subordinate midwifery, to limit its work to peripheral tasks and in some instances to ban it from legal practice (Willis 1989, Bourgeault *et al.* 2004). This understanding of professional relationships draws, implicitly or explicitly, on the neo-Weberian concept of social closure and patriarchy (Abbott 1988, Witz 1992). This (male) power-centric view of the professions has been challenged by some who argue that professional boundaries are sometimes successfully contested, though not always in the manner that allied professionals themselves hoped for (Allen 1997, 2000). Thus, the power of medicine is not all-supreme over midwives' day-to-day work (Lay 2000).

Consumer interests
The final element in our framework is consideration of consumer impact on maternity care policy. The literature on consumer groups describes the role of activists and ordinary women in pressuring for change in the organisation of maternity systems. Study of the dynamics of collective challenges to policy and practice requires sensitivity to the institutional context of policymaking and service provision (Bourgeault, DeClercq and Sandall 2001). Our case studies show how social movements are socially-situated and vary in intensity and impact on maternity care systems and the position of midwives.

Cross-national comparison of maternity care systems

United Kingdom
At different points, a welfare-centred approach has dictated the organisation of maternity care services in the UK, either in consort with, or against the interests of the medical profession. Moreover, the maternity care arrangements across the historical period point to both the gendered and interactive nature of the maternity care domain.

In the UK, pronatalist 'politics of motherhood' resulted in the creation of social and health programmes for mothers early in the 20th century, though midwives initially had to fight hard to be included in state provision. By the 1930s, two models of care were predominant, a consultant-based hospital system and a community-based maternal and child health system. The 1936 *Midwives Act* and the National Maternity Service was seen as part of the solution to high levels of maternal mortality. The goal was to provide a national salaried community-based midwifery service, including antenatal and postnatal care, home birth and general practitioner (GP) back up.

The creation of the National Health Service (NHS) in 1948 gave women access to free maternity care for the first time and strengthened consultant-based hospital services. By the 1970s, public pressure for hospital-based services had grown, GPs had taken over the provision of care from municipal midwife-run clinics, and obstetricians became increasingly involved in 'normal' childbirth; in 1970 the government recommended hospitalisation for all births (DHSS 1970). Although midwives continued to be the primary attendant at the majority of hospital births, their role became fragmented as a result of new technologies and limits on jurisdiction and scope of practice. Thus, we see change from state support for autonomous midwives to medical dominance over their mandate and practice.

The main consumer organisations concerned with maternity care – the National Childbirth Trust, Association for Improvements in Maternity Care, and Maternity Alliance – began to challenge such medical dominance. In the 1980s research evidence also began to play an important role in legitimising these concerns, resulting in an alliance between consumers, the Royal College of Midwives (RCM) and the interest groups identified above (Sandall 1996). This coalescence created a 'window of opportunity' to introduce policy change (Kingdon 1995).

As technological interventions increased in the 1990s, consumer organisations came to play an even more prominent role in the childbirth debate, resulting in media and parliamentary attention. Medical dominance was questioned and the welfare state expressed support for less-powerful female actors – midwives and consumers. One result was a parliamentary enquiry to assess the most appropriate and cost effective use of professional expertise in the maternity domain. Where previous inquiries had tended to focus on mortality rates, this time normal pregnancy and birth were the centre of attention. The *Winterton Report* (House of Commons 1992) and the government's response, *Changing Childbirth* (DH 1993), critically reassessed the roles of health professionals and used the views and experiences of women in the creation of policy recommendations. Both reports concluded that there should be less focus on mortality rates as the major outcome measure and recommended a move towards a 'woman-centred' approach that offered women choice in place, type of service and 'continuity of care'. According to Annandale and Clark (1996: 424), this shift in thinking allowed 'the integration of feminist interests, the grass roots feelings of women, [and] the heart of midwifery philosophy' to be realised in practice.

It is also possible to see these recommendations as part of a broader reform effort emphasising cost-efficiency. For example, *Changing Childbirth* encouraged choice, personal continuity and control, all neo-liberal ideas that link with welfare pluralism and 'lean production'. It ignored the wider range of social and environmental effects on health highlighted in the *Winterton Report* in favour of a strategy that treated maternity care as a vehicle for the expression of consumer values (Streetly 1994). Viewed through this lens, policy-makers used the argument of women's interests to pursue particular management aims. Not surprisingly, the implementation of the *Changing Childbirth* policy has been patchy. Pilot schemes were not mainstreamed and large variations remain in provision of care. Childbirth interventions continue to increase (ONS 2003) and concerns exist about the retention of midwives.

More recently, Parliamentary Reports on maternity care have expressed concerns yet again regarding choice, rising caesarean section rates and inequalities in care (e.g. House of Commons 2003c). It is noteworthy that the new government policy separates out services for mothers from other women's health services (DH 2004). Legitimated by increasing evidence on the science of foetal programming (Godfery 2001), it argues that 'improving the health and welfare of mothers and their children is the surest way to a healthier nation' (DH 2004). In essence, the new policy for maternity care espouses a public health model of midwifery. Such a model explicitly encourages an expanded midwifery jurisdiction in areas of care previously provided by GPs and health visitors and is supported by the increase in numbers of maternity assistants. This scenario is less the result of professional negotiations and consumer stakeholder power, and more that of the modernisation agenda in the NHS where hospital reconfiguration and refocusing of the work of GPs and obstetricians have left a vacant occupational space (Abbott 1988).

Finland

It is estimated that nearly 80 per cent of births in Finland are attended by midwives (Gissler 2005). Official statistics are not available as the publicly-provided birth care in Finland relies on midwives as sole attendants for all normal vaginal births and also in case of minor complications. Despite the focal role of midwives in birth care, home birth accounts for less than one per cent (Viisainen 2001: 1110). While Finnish women have begun to challenge the dominance of the medical definition of birth in the maternity care system, midwives have not joined in and articulated jurisdictional claims over 'non-medical normal birth', as has been the case in the UK. What are the dynamics of the Finnish approach to maternity care that produce this particular situation?

Finnish society is shaped by a state-centred strategy of making social and health policy, of which the formation of maternity care is but one example. Consensus-oriented negotiations, aimed at accommodating diverse class and gender interests, are an established feature of the political culture. The Finnish state has been both an engine of welfare policy and a site of negotiations, to which professional groups and other societal actors have been tied. Rather than 'women-friendly' in the simple meaning of the term, the relative social equality between men and women and social classes has been achieved through complex negotiations among Finnish trade unions, stakeholders in the economy and finance, political parties and interest groups. The accommodation of diverse societal interests in the organisation of healthcare has served as a counterweight to the position of the medical profession. Yet this has not necessarily been to the advantage of autonomous midwifery or alternative forms of maternity care.

In 1944, the Parliament adopted legislation that introduced free maternity and infant care. For a nation that was still waging war, such investment represented hope for the future welfare of the population. Private-practice municipal physicians lost ground to midwives, who gained a state mandate to provide both social and healthcare services to women during pregnancy and birth. For mothers, the new public health services represented a social obligation to seek care as much as a social right to receive it (Wrede 2001: 256). Emphasis on women's rights as citizens, not merely as mothers, occurred much later, with the introduction of municipal day care services in 1973 – a policy associated with women's equality in the labour market (Bergman 2002).

Free maternity care in the long run led to increasing hospitalisation of birth where obstetricians held sway. At the same time, the role of the municipal midwife became narrowed to a focus on preventive care, a situation that led local health policy makers to question the need for their high level of competence. Lack of support from municipal decision makers, together with plans to increase physician involvement in primary care, resulted in the exclusion of municipal midwives from the new model of first-line care, enacted through the *Public Health Act* of 1972. A two-tier maternity care system was designed. Public health nurses, in collaboration with primary care physicians, were awarded responsibility for prenatal care during normal pregnancies; hospital-based midwives, in collaboration with obstetricians, were responsible for pregnancy complications and all births. The Act thus split the midwifery profession and fragmented maternity services. Paradoxically, the overall aim of Finnish healthcare policy at the time – to promote primary care – ended up paving the way for the increased medicalisation of pregnancy and childbirth (Hemminki *et al.* 1990).

The position of the state as a central site of negotiations between societal interests was reinforced as policy expanded to new areas in the 1970s. The creation of the municipal primary care system in 1972, for instance, was based on the implementation of social democratic ideals of equity and equal access. The primary care reform occurred at the same time as feminists raised the issue of public day care as the dominant 'women's issue' (Bergman 2002). Yet support for midwifery did not gain such a status; feminists instead criticised the patriarchal notion of the family inherent in the organisation of maternity care. Health policy makers responded by encouraging fathers to become more active parents in the early years of their child's life (Wrede 2001). A short leave for fathers in association with the birth of a child was introduced in 1978, and in 1980 the concept of 'parental leave' was created. In this reform the former maternity leave was lengthened and divided into two periods of which the first one was the actual maternity leave, intended for women after the delivery, whereas the parents were free to choose which of the parents would stay at home during the later, longer period. In addition to these arrangements, maternity care providers were encouraged to involve fathers throughout the childbearing process (Wrede 2001). Universalistic welfare policy thus served to politicise healthcare delivery and undermined the position of professional interest groups (such as midwives) in favour of state authorities.

All in all, subsequent reforms of primary care centres, including the recent neo-liberal reforms aiming at improving flexibility and cost-effectiveness, continue to maintain the two-tier character of the maternity care system. The issue of women-centred maternity care provided by midwives has not been formulated in feminist terms in the same way as it has in the UK (above) and Canada (see below).

In a comparative perspective, the lack of feminist debate around maternity care calls for an explanation. Our perspective suggests a few reasons. First, in a country where all women have had access to an extensive maternity service for practically no cost, and where infant and maternal outcomes are among the best in the world, other goals such as public day care, workers' rights and equality in the family have topped the feminist agenda. Furthermore, the Finnish version of social equality resulted in feminists emphasising similarities in the interests of men and women (Bergman 2002). Yet an emphasis on women's role as childbearers has been difficult to combine with the social democratic rhetoric of sameness. In this context, a demand for choice in birth care has become framed as an individualistic, even elitist issue. Even though at times vocal, the only longstanding network of birth activists, *Aktiivinen synnytys* (Active birth), is small, with a revolving membership of roughly 400 members. Founded in the mid-1980s, the network has targeted 'birth culture' rather than social policy. In the few instances that birth activists have engaged in policy (Viisainen 2001: 1/10), they have lobbied against decisions to close down particular maternity units.

Our finding of weak feminist and general public support for an autonomous midwifery profession suggests that the politics of gender sameness has a significant impact on the organisation of maternity care services in Finland. In such an equality-driven society, reproduction and the organisation of maternity care remain oddly under-politicised.

The Netherlands

The Netherlands is well known among maternity care researchers and women's health activists for its unique system of maternity care. Unlike the other three countries analysed in this chapter, the Netherlands actively promotes birth at home under the care of primary caregivers – midwives and GPs. Consequently, more than 30 per cent of births occur at home, the majority attended by midwives. The Dutch state has a history of preserving autonomous midwifery and birth at home through: (1) laws and regulations that give preference to midwifery care; (2) state support for the education of midwives and for the conditions of midwife practice; and (3) funding of research that demonstrates the safety and efficacy of midwife-attended home birth. All three suggest a relatively strong welfare state as the dominant stakeholder, similar in some respects to its Finnish counterpart discussed above, but quite different in its women-friendly as opposed to gender neutral focus.

State support for midwifery was established in 19th century legislation that defined the practice of medicine. In both the 1818 *Health Act* and the 1865 *Medical Act*, midwifery was defined as a 'medical' profession and given a well-defined sphere of practice. When the modern system of national health insurance was created in the 1940s, midwives were given preference as women's first choice provider and this arrangement persists to this day. Thus, a healthy Dutch woman opting for a home birth might not see a doctor at all during her pregnancy and birth. Midwives provide antenatal and postnatal care in the community and attend home births and short-stay hospital deliveries.

The Dutch have a national health insurance system, with a mixture of collective health insurance organisations (so-called Sick Funds) and commercial health insurers. A government organised committee – which includes representatives from employers, unions, caregivers (of all sorts), hospitals, patient organisations, insurance companies and government officials – makes important decisions about what sort of care will be offered, by whom, and for what fee. This centralised control has been used to promote midwifery: women who expect a normal ('physiological') birth – as defined by an Obstetric Indications List – must receive their care from either a midwife or a GP. Specialist care can be used only when complications are present. This legislation limits competition, and more importantly, guarantees business for midwives.

The government also keeps a close eye on the conditions of midwife practice, adjusting policy in an effort to keep an adequate supply of midwives. In response to a recent shortage of midwives, the Minister of Health authorised both the creation of a new school of midwifery and an increase in the number of new students (from 120 in the 1990s to 220 in 2003–06). In conjunction with this increased flow of students into the profession, the government also improved the income of working midwives (De Vries 2005: 94–141). The desire to support midwives also prompted policy changes that increased the number of *kraamverzorgende* (post-partum caregivers) who assist midwives during the birth and support mothers and families in the postpartum period. The work of the *kraamverzorgende* makes the task of midwives easier during the postnatal period. As part of a conservative family-centred welfare state perspective, rather than enacting parental policies which would give fathers more time to spend with their family (as in the Finnish case), the Dutch state has spent its resources on the *kraamverzorgende*, a social support person/ husband substitute available to the woman during and after childbirth.

A final source of government support for midwifery is the funding of research on the Dutch way of birth, research that guides policy decisions for the profession, including the working conditions of midwives (Bakker *et al.* 1996) and the effectiveness of the maternity care system (Wiegers 1997: 48). The latter concluded that, 'perinatal outcome was significantly better for planned home birth than for planned hospital births', for both first-time mothers and those with previous children.

Because of strong government support, medical-technological developments in maternity care have not had a major impact on midwifery. Abraham-van der Mark (1993) summarised:

> Although midwives lost ground in the twentieth century in other Western countries, Dutch midwifery was characterised by growing professionalisation: midwives' qualifications were increased, standards for recruitment and training were made more rigorous, and their organisation gained power.

The well-established position of Dutch midwives has resulted in a virtually non-existent consumer movement concerned with choice of maternity provider and birth place. The particular type of gendered policies of the Dutch state supported pregnant women's interest most of the time when it supported midwives. On occasion, when these two sets of interests were not parallel or when the state failed to support midwifery, short-lived consumer groups sprang up (and rapidly disappeared): in the mid-1980s *Beter Bevallen* (with the double meaning of 'better delivery' and 'more pleasing') was organised, and in the late 1990s a group of parents and midwives created the *Stichting Perinatale Zorg en Consumenten* (Foundation for Perinatal Care and Consumers). The latter was a reaction to the declining home birth rate and the subsequent risk of the elimination of birthplace choice for Dutch women. It was this latter group, coupled with midwives' political activism, that encouraged the government to take seriously the complaints of midwives about their working conditions. The subsequent policies – increased salary and reduced caseloads – have been responsible for both stabilising the rate of homebirths and bolstering the midwifery profession (De Vries 2005).

Why has the Dutch government consistently supported midwifery and protected home birth? The answer lies in a mix of cultural ideas held by the general public and structural features of medicine. The Dutch are noted for their distinctive ideas about home, family, the efficacy of medicine and *zuinigheid* (thriftiness). If Finnish welfare policy in the 1970s had a gender-neutral emphasis, in the Netherlands an ideology of men and women's complementary roles in the family and in society has informed maternity care policy. This conservative welfare state ideology has shaped policies that have reinforced a view of birth that is home-based and family-centred (van Teijlingen 2003). Thus, as long as the policies are (seen to be) family-centred there is little need for consumers to put pressure on the political decision-making process.

Dutch ideas about *zuinigheid* imply more than just being cautious about spending money; it is also possible to be thrifty in one's response to other aspects of life. Dutch *zuinigheid* is associated with a very pragmatic approach to all social policy: decisions about everything from drug use, to euthanasia, to childbirth are made only after careful study of different approaches (De Vries 2005). While other countries were abandoning homebirth and mid-wifery for less than scientific reasons (Tew 1995), the Dutch examined these practices and found them to be safe and effective as well as family-centred. The Netherlands is also distinctive in the slow development of the specialty of obstetrics. Although Dutch obstetrics is now the equal of any in the world, it was not always so, a fact that allowed a strong profession of midwifery to be established and continue to exist autonomously into the 21st century.

Canada
Because Canada is a latecomer in designing a maternity care system that includes midwifery, the negotiated and gendered nature of that system domain reflects a more contemporary period. Even as late as the early 1970s when the Canadian government began to fund maternal services for all pregnant women, the system was created exclusively for medical and hospital approaches to mater-nity care (Benoit and Heitlinger 1998). This was but a more recent example of the exclusionary patriarchal social closure strategies (Witz 1992) that had been directed towards Canadian midwives since the turn of the 20th century.

First, the *Medical Practitioners Acts* of most provincial health ministries (the agencies that oversee health service provision) restricted the perform-ance of 'maternity services' almost exclusively to licensed members of the College of Physicians and Surgeons until the early 1990s. Second, public funds for maternity care services have been available only for hospital services since the late 1950s and specific activities performed by licensed physicians since the early 1970s. So-called 'alternative' health services, such as midwifery, remained, until recently, uninsured in provincial healthcare plans. As a result, midwifery services have not been available, unless birthing women were able to pay privately and access them locally. Private-practice midwives also had to work outside the official healthcare system and formal healthcare settings (Benoit and Heitlinger 1998). Thus, through its systems of provincial health insurance and licensed Medical Acts, the Canadian welfare state enshrined medical dominance over the country's maternity care services.

In recent years, much to the chagrin of the medical profession, some provincial governments have enacted policies integrating certified trained midwives and a homebirth option into the formal healthcare system – either with protective licensing, public funding for midwifery services, or both. This includes midwifery care for the ante, intra and postpartum periods. In other words, the new Canadian midwives are legally able and in some provinces required to attend a certain number of both home and hospital births annually, while their physician counterparts are banned from attend-ing homebirths by their professional licensing body.

Two intertwined pressures on the Canadian welfare state – cutbacks in federal health funds to the provinces, and an integrated consumer/midwifery social movement demanding greater choice in childbirth – helped to bring about this turn (Bourgeault *et al.* 2004). This is similar to Britain where the interests of a retrenched welfare state and consumers came together but, as noted above, was largely absent in the Finnish and Dutch cases.

The success of midwifery and consumer groups in shaping the health policy agenda of particular Canadian provinces resulted from several factors. The case of Ontario, the first province to legalise midwifery, is instructive. First, midwifery organisations and consumer groups garnered support from key actors within the provincial government who offered significant support for the integration of midwifery into the healthcare system. Support for birth activists was visible in the establishment and funding of midwifery policy committees, the purposive appointment of persons with pro-midwifery sentiments to these committees, and the government's response to these committee's recommendations (Bourgeault 2005). As noted in the British case, state support for the midwifery initiative was forthcoming for two main reasons. First, midwifery was regarded as cost-effective, which suited the government's rationalisation of the health-care system. Secondly, it made the government appear as 'progressive' by promoting women's rights. It is also not insignificant that the Ontario Ministers of Health throughout the midwifery integration process were strong feminists who were personally supportive of midwifery as a woman-friendly initiative.

Midwives can now practise legally in five Canadian provinces and midwives' services are now included as part of provincially-funded public healthcare services in four of these, while similar legislation is underway in several other provinces and territories. The midwifery integration process in Canada offers an example of a maternity care policy initiative fuelled by organised consumer support but also attractive to governments pursuing particular agendas. At the same time it is important to note that these changes have occurred against a backdrop of an overall reduction in the obstetric work-force in the 1980s that created a vacant jurisdictional space in maternity care division of labour. Many physicians – both GPs and specialists – have abandoned the practice of maternity care because of significantly higher malpractice insurance fees and the demands of obstetric care on one's personal life. In 1983, 68 per cent of family physicians attended births; by 1995 this number was reduced to just 32 per cent (Reid *et al.* 2001).

Canadian midwifery care, however, remains an alternative for a small albeit growing group of women. Registered midwives – like other healthcare providers – for the most part work in or near large urban centres. Their numbers are still small; currently they attend only two per cent of births nationally. So although the recent introduction of midwifery was a major consumer-driven and state-supported maternity care policy initiative, it has to date only had a small ripple effect on the national healthcare system. Indeed, the presence of midwives may cause further decline in GP services, leaving birthing women without any choice of health provider. Moreover, while the boundaries between midwives and physicians are being successfully challenged and re-negotiated, the latter still wield enormous power over the future direction that the country's maternity care system will take.

Discussion

Comparative analysis of the roles and responsibilities of midwives reveals significant cross-national variation in the social organisation of maternity care systems. The early English language sociological research largely focused on North America, where midwifery had been marginalised or excluded from the health division of labour and subsequently idealised as an authentic 'alternative' to medical control (Lay 2000). Research on the other side of the Atlantic reveals that midwifery has had other historical paths and holds multiple meanings today, resulting from conflict and negotiation at the macro, meso and micro levels of healthcare organisation. Our 'decentred method' makes it possible to shed light on how different welfare states organised their respective maternity care services, how boundary disputes in the health division of labour were contested, negotiated and renegotiated (or not), when and where consumer groups mobilised around maternity care issues and variation in the gendered dynamics across all three levels of analysis. Welfare regimes are revealed as sites of political struggles that involve complex and interactive relations between the state itself, the historically male-dominated professions, the contested female occupation of midwifery and the women who use maternity services.

World War II and its aftermath was associated with the reshaping of the health-care systems in our European cases, generating reforms that emphasised state responsibility for the provision of healthcare. In Europe, this fact overrode other emerging cultural differences in welfare regimes. In Canada, the war lacked such collective impetus. Notice, for example, how the UK – following the Northern European tradition of integrating midwifery into the official health service – has kept midwifery alive, while in Canada, until recently, midwifery suffered under the North American approach that favoured market-based physicians as providers of maternity care. Recent moves by provincial governments in Canada to support midwifery have had as much to do with efforts towards cost containment and the rational use of resources as to a 'women-friendly' welfare state approach to re-organising maternity care. In terms of the theory of professions, Canada serves as an example of a system in which a 'vacant jurisdiction' (Abbott 1988) is emerging in maternity care, because more and more physicians are choosing not to offer the services, and because their opposition to midwives is not as strong as before. Midwives have been seen by some governments as a cost-effective way to help fill this void.

But this is not the full story. Our analysis also highlights the multiple ways in which those who use maternity care services have made organised efforts to change the way care at birth is given. Again in the Canadian case as well as the UK, there have been active birth movements organised as consumer groups that support midwifery as an ideal means of achieving women-centred care. Consumer organisations have played an increasing role in the debate around changing maternity care practices, resulting in media as well as government interest (Bourgeault, DeClercq and Sandall 2001). Birth activists have hardly played a role as pressure groups in Finland and the Netherlands, however, as countries where the government continues to intervene in maternity care in ways that limit market principles. In the Finnish case, maternity care policy is focused on equality between parents during the childbearing period, which undermines collaboration between midwives and birthing women, and results in a maternity care system splintered between primary care providers and hospital-based obstetricians and midwives. The Dutch welfare state, while also adopting a lead role over the organisation of maternity care, has historically favoured a more gender-specific and in many ways more woman-friendly approach, based on an ideology of complementary roles between men and women. By way of contrast, the UK and Canada have emphasised market principles, making women important as consumers, which frames the issue of maternity care in a more individualistic way.

In all four health systems midwifery is constantly redefined in relation to medicine, requiring a set of theoretical concepts that draw from all three theories mentioned in the introduction of this chapter. Some have argued that the overriding logic of health systems in high-income countries is the logic of the medical professions, even when 'pockets' for other logics emerge from time to time (Griffiths 2003). The idea of woman-centred care has made inroads into the organisation of maternity care, but, lacking the support of arguments such as a possibility for cost containment, it could not compete with the logic of medicine. Similar to work by Allen (1997) on negotiation over the nursing-medical boundary, our analysis suggests that the logic of medical dominance can nevertheless be successfully challenged under certain circumstances, for example when maternity care is viewed as a social entitlement by the welfare state and where activists see midwifery as a vehicle to achieve accessible care. At present, however, the Netherlands is the only welfare state that truly supports midwifery as the favoured woman-centred solution for the provision of maternity care, supported interestingly enough, by a fluid consensus among state policy makers, GPs, obstetricians, midwives and consumers at large, in a structure that equalises the power of each partner to negotiate. The design of maternity services in the other three countries remains an outcome of often-times competing welfare state interests, professional boundary struggles and changing consumer interests surrounding pregnancy and childbirth.

Acknowledgements

Initial funding for the research project on which this paper is based was provided by the Council for European Studies (Columbia University, New York), the Academy of Finland, the Finnish private foundation, Stiftelsens för Åbo Akademi forskningsinstitut, and the Royal Dutch Organization of Midwives. Additional sources of funding for the individual authors are: Cecilia Benoit: the National Network on Environments and Women's Health; Sirpa Wrede: the Academy of Finland project,

'New Dynamics of Professionalism within Caring Occupations' (# 202556); Ivy Bourgeault: Social Sciences and Humanities Research Council of Canada, the National Health Research and Development Program of Health Canada, the Canadian Institute for Health Research; Jane Sandall: NHSE London Region, 'The Changing Role of the Midwife in Relation to the Consultant Obstetrician and GP' (#OM906); Edwin van Teijlingen: SHEFC, Edinburgh; Raymond De Vries: the US National Institutes of Health (Grant number: F06-TWO1954), NIVEL (Netherlands Institute for Health Care Research) and the Catharina Schrader Stichting.

Notes

1 The United States was purposely excluded because it is an outlier in regard to many of these health system features.
2 In this context, we use 'touchstone' to refer to midwifery as the key for understanding the variation in maternity care systems across these four developed welfare states.

References

Abbott, A. (1988) *The System of Professions. An Essay on the Division of Expert Labor*. Chicago: The University of Chicago Press.
Abraham-Van der Mark, E. (ed.) (1993) *Successful Home Birth and Midwifery: the Dutch Model*. Westport, Connecticut: Bergin and Garvey.
Allen, D. (2000) Doing occupational demarcation, *Journal of Contemporary Ethnography*, 29, 3, 326–57.
Allen, D. (1997) The nursing-medical boundary: a negotiated order? *Sociology of Health and Illness*, 19, 4, 498–520.
Anderson, B. (1983) *Imagined Communities: Reflections on the Origin and Spread of Nationalism*. London: Verso.
Annandale, E.C. and Clark, J. (1996) What is gender? Feminist theory and the sociology of human reproduction, *Sociology of Health and Illness*, 18, 1, 17–44.
Bakker, R.H.C., Groenewegen, P.P., Jabaaij, L., Meijer, W.J., Sixma, H.J. and De Veer, A.J.E. (1996) 'Burnout' among Dutch midwives, *Midwifery*, 12, 174–81.
Benoit, C. and Heltlinger, A. (1998) Women's health caring work in comparative perspective: Canada, Sweden and Czechoslovakia/Czech Republic as case examples, *Social Science and Medicine*, 47, 8, 1101–11.
Bergman, S. (2002) *The Politics of Feminism: Autonomous Feminist Movements in Finland and West Germany from the 1960s to the 1980s*. Åbo: Åbo Akademi University Press.
Bourgeault, I.L. (2005) *Push! The Struggle to Integrate Midwifery*. McGill: Queen's University Press.
Bourgeault, I., DeClercq, E. and Sandall, J. (2001) Changing birth: interest groups and maternity care policy. In DeVries, R., Benoit, C., van Teijlingen, E. *et al.* (eds) *Birth by Design: Pregnancy, Maternity Care, and Midwifery in North America and Europe*. New York: Routledge.
Bourgeault, I., Benoit, C. and Davis-Floyd, R. (eds) (2004) *Reconceiving Midwifery: the New Canadian Model of Care*. Montreal/Kingston: Mc-Gill-Queen's University Press.

Coburn, D. (2001) Health, health care, and neo liberalism. In Armstrong, P., Armstrong, H. and Coburn, D. (2001) *Unhealthy Times: Political Economy Perspectives on Health and Care.* Toronto: Oxford University Press.

Davies, C. (2003) Some of our concepts are missing: reflections on the absence of a sociology of organisation in 'Sociology of Health and Illness', *Sociology of Health and Illness*, 25, 172–90.

DHSS (1970) *Report of the Sub-committee on Domiciliary and Maternity Bed Needs* (Chairman, Sir John Peel). London, HMSO.

Department of Health (1993) *Changing Childbirth, Part 1: Report of the Expert Maternity Group.* London, HMSO.

Department of Health (2004) *National Service Framework for Children, Young People and Maternity Services, Maternity.* London, Department of Health.

De Vries, R. (2005) *A Leasing Birth: Midwifery and Maternity Care in the Netherlands.* Philadelphia, PA: Temple University Press.

De Vries, R., Benoit, C., van Teijlingen, E. and Wrede, S. (eds) (2001) *Birth by Design: Pregnancy, Maternity Care and Midwifery in North America and Europe.* New York: Routledge.

Esping-Andersen, G. (1990) *The Three Worlds of Welfare Capitalism.* Cambridge: Polity Press.

Gissler, Mika (2005) Personal communication from the Manager responsible for the Medical Birth Register compiled by the National Research and Development Centre for Welfare and Health, 18.03.2005.

Godfrey, K.M. and Barker, D.J. (2001) Fetal programming and adult health, *Public Health Nutrition*, April, 4, 28, 611–24.

Griffiths, L. (2003) Making connections: studies of the social organisation of health care, *Sociology of Health and Illness*, 25, Silver Anniversary Issue, 155–71.

Hemminki, E., Malin, M. and Kojo-Austin, H. (1990) Prenatal care in Finland: from primary to tertiary care, *International Journal of Health Services*, 20, 221–32.

House of Commons (1992) *The Health Committee Second Report: Maternity Services*, Volume 1, Chair N. Winterton. London: HMSO.

House of Commons Health Committee (2003) *Choice in Maternity Services, Ninth Report of Session 2002–03*, Volume 1. London: The Stationery Office.

Kingdon, J. (1995) *Agendas, Alternatives and Public Policies.* 2nd Edition. New York: Harper Collins.

Korpi, W. (2003) Welfare – state regress in Western Europe: politics, institutions, globalization, and Europeanization, *Annual Review of Sociology*, 29, 1, 589–609.

Lay, M.M. (2000) *The Rhetoric of Midwifery. Gender, Knowledge and Power.* New Brunswick: Rutgers University Press.

Lewis, J. (1992) (ed.) *Women and Social Policies in Europe. Work, Family and the State.* Aldershot: Edward Elgar.

ONS (2003) NHS *Maternity Statistics, England: 2001–02.* London: Government Statistical Service.

Reid, A., Grava-Gubins, I. and Carroll, J. (2001) *National Family Physician Workforce Survey (weighted data), the Janus Project.* Mississauga: College of Family Physicians of Canada.

Ruggie, M. (1984) *The State and Working Women.* Princeton, NJ: Princeton University Press.

Sandall, J. (1996) Continuity of midwifery care in England: a new professional project? *Gender, Work and Organisation*, 3, 4, 215–26.

Streetly, A. (1994) Maternity care in the 1990s. *Health For All 2000 News*, 26, 14–15.

Teijlingen van, E. (2003) Dutch midwives: the difference between image and reality. In Earle, S. and Letherby, G. (eds) *Gender, Identity and Reproduction: Social Perspectives*. London: Palgrave.

Tew, M. (1995) *Safer Childbirth? A Critical History of Maternity Care*. London: Chapman and Hall.

Viisainen, K. (2001) Negotiating control and meaning: home birth as a self-constructed choice in Finland. *Social Science and Medicine*, 52, 1109–21.

Wiegers, T. (1997) Home or hospital birth: a prospective study of midwifery care in the Netherlands. Ph. D. Thesis Leiden University. NIVEL, Utrecht.

Willis, Evan (1989) *Medical Dominance: the Division of Labour in the Australian Health Care System* (2nd Edition) Sydney: George Allen and Unwin.

Witz, A. (1992) *Professions and Patriarchy*. London: Routledge.

Wrede, S. (2001) *Decentering Care for Mothers. The Politics of Midwifery and the Design of Finnish Maternity Services*. Åbo: Åbo Akademi University Press.

Chapter 4

Managerialism in the Australian public health sector: towards the hyper-rationalisation of professional bureaucracies

John Germov

Introduction: managerialism and professional bureaucracies

Professional bureaucracies, particularly the organisation of hospitals, have long been recognised as deviating substantially from Weber's (1968/1921) ideal type[1]. They consist of professional workers who require a high level of autonomy in decision-making because of the specialist, complex and indeterminate nature of their work (Mintzberg 1979). Hospitals have also been conceptualised as a 'negotiated order' (Strauss *et al.* 1963), involving the dual power structures of bureaucratic rules (formal rationality) and professionalism (substantive rationality) (Turner 1995, Freidson 1994). Since the 1980s, a range of managerial strategies such as Management by Objectives (MbO), Total Quality Management (TQM), and 'best practice' have been introduced into public bureaucracies – a process commonly referred to as managerialism. A significant literature has examined the impact of managerialism on professional work in health bureaucracies, particularly on medicine and nursing, with an emerging consensus that the work of health professionals (to greater and lesser degrees) is subject to enhanced rationalisation and intensification (*cf.* Allsop and Saks 2003, Sheaff *et al.* 2003, Flynn 2002, Harrison and Dowswell 2002, Harrison and Ahmad 2000, Pollitt and Bouckaert 2000, Exworthy and Halford 1999, Dent 1998, Coburn *et al.* 1997, Walby and Greenwell 1994, Freidson 1994).

This chapter discusses empirical findings from 11 qualitative case studies of the introduction of best practice managerial strategies in the Australian public health sector. The case studies are drawn from the Australian Federal Government funded Best Practice in the Health Sector (BPHS) programme. The programme provided funding on a competitive tender basis for small-scale projects to develop and disseminate best practice models of work organisation in healthcare settings (Germov 2002, Germov *et al.* 1998, DHSH n.d.). The voluntary nature of the BPHS programme represents a novel example of managerialism because participants chose to introduce best practice managerial reforms into their workplace.

The concept of best practice involves a prescriptive list of work practices identified among a number of successful private sector companies, with the key idea being that all organisations should adopt the practices of 'the best' to improve their performance (Dertouzos *et al.* 1989, Camp 1989). It is based on the assumption that generic laws of management can be identified and that these are appropriate to all organisations, echoing the logic of Taylor's (1947/1911) work on 'scientific management'. Bearing a similarity to the 'quality movement' literature (see Peters and Waterman 1982), best practice can be defined as involving: fewer hierarchical organisational structures, less teamwork, less of a commitment to continuous improvement and learning, a focus on internal and external customers, less benchmarking and use of performance measurement systems (DIR/AMC 1992).

The Australian experience of managerialism
While managerialism has occurred in many countries, there remain significant national variations on the managerial trend (Pollitt and Bouckaert 2000). To set the context for the present study, this section provides a brief overview of the Australian experience of managerialism relevant to the health sector.

In the 1980s, managerialism in Australia primarily concerned the reorientation of the central agencies of the public sector towards programme structures, programme budgeting and performance measurement. These reforms were complemented by the privatisation of some public enterprises, user-pays for some public services, and a shift from industrial awards towards enterprise bargaining (APS 2003). By the early 1990s, Australian managerial reforms increasingly focused on the service delivery agencies of public bureaucracies by drawing upon managerial techniques such as TQM and best practice, and subjecting service delivery to market rationality through competition and contracting-out (as promulgated by Osborne and Gaebler 1992). The main developments have been competitive tendering for support services (such as cleaning, printing and information technology), the creation of internal markets among public utilities, and purchaser-provider arrangements for the delivery of employment services. The Australian experience of managerialism can usefully be summarised as exemplifying Hoggett's (1996) notion of 'centralised-decentralisation'; whereby government has centralised strategic control over public sector policy, performance targets and resources, while operational decision-making and service delivery has been decentralised (see O'Brien and Fairbrother 2000, O'Brien and O'Donnell 2000) – a process that Rose and Miller (1992) refer to as 'government at a distance'.

It is difficult to give a concise synopsis of the managerial reforms introduced into the Australian health system because of the country's federal political structure. Australia has three tiers of government – a federal (national) government, six state and two territory governments, and local governments – all of which share responsibility for health services. The federal government is the key policy-making body and, while not directly providing health services, has a major influence on the health system through its disbursement of general taxation funds to the states and territories, particularly through its funding of Medicare. Medicare is a universal public health insurance scheme that provides free treatment in public hospitals and free or subsidised general practitioner services. State and territory governments are responsible for the delivery of acute public hospital and psychiatric services, residential aged care and rehabilitation services, community health services, the regulation of health professionals and environmental health programmes (in conjunction with local governments) (Belcher 2005). Across the health system, the common managerial reforms introduced have involved quality accreditation systems, performance-based funding (casemix and diagnosis-related groups), and a range of targeted programmes to facilitate the adoption of specific managerial strategies such as TQM and best practice.

There is a voluminous literature debating the impact of managerialism on the Australian public sector, which has mostly concentrated on concerns about the marginalisation of equity and quality service delivery in the face of productivity increases achieved by 'doing more with less' (see Hughes 2003, Stanton 2002, O'Donnell and Shields 2002, O'Brien and O'Donnell 2002, Beattie 2000, O'Donnell et al. 2001, O'Brien and Fairbrother 2000, O'Donnell and O'Brien 2000, Halligan 1997, Braithwaite 1997, Considine and Painter 1997, Yeatman 1994, Keating 1990, Considine 1990, Paterson 1988, Bryson 1987). In particular, studies have found that budgetary concerns, performance pay and perceptions of nepotism in performance ratings of staff have undermined staff morale and team work (O'Donnell and Shields 2002, O'Brien and O'Donnell 2000). Performance-based funding of hospitals has raised concerns over the premature release of patients to keep costs down (discharging patients 'quicker and sicker'), with hospitals shifting responsibility for continued care on to the community (Draper 1992, Braithwaite and Hindle 1998, Lin and Duckett 1997). Moreover, managerial reforms have been associated with increased levels of stress and work intensification among health professionals (see Weekes et al. 2001, Allan 1998).

The underlying logic of managerialism in professional (public) bureaucracies: a neo-Weberian contribution

The question remains as to whether there is an underlying logic to the way managerialism has affected professional bureaucracies, particularly in terms of how the conduct of professional autonomy is governed. Ferlie et al. (1996) suggest that managerialism is resulting in a hybrid organisational form based on a mix of public and private sector managerial strategies. Walby and Greenwell (1994) in their UK study of managerial impacts on nurses found both neo-Fordist and post-Fordist forms of work organisation co-existed in constant tension. Light's (2000: 203) pluralistic notion of countervailing powers highlights the mediated nature of professional autonomy, whereby stakeholders 'pursue their interests resulting in power struggles between professional and organisational actors that can lead to "new organizational forms"'. Courpasson (2000) uses the term 'soft bureaucracy' to convey how professional autonomy is regulated through a hybrid of decentralised and bureaucratic authority structures (see also Sheaff et al. 2003).

Dean (1999), using a governmentality perspective, suggests that new managerial strategies comprise 'technologies of performance' that:

subsume the substantive domains of expertise (of the doctor, the nurse, the social worker, the school principal, the professor) to new formal calculative regimes . . . Here, the devolution of budgets, the setting of performance indicators . . . are all more or less technical means for locking the moral and political requirements of the shaping of conduct into the optimization of performance (1999: 169).

From a similar perspective, Flynn (2002: 169) examined clinical governance in the UK and found 'a mixture of rationalities and strategies designed to establish and codify explicit standards and to achieve a rigorous method of performance evaluation through the co-optation of medical professionals in ways which give some semblance of delegated autonomy'.

This chapter presents an alternative conceptualisation of the impact of managerialism on professional (public) bureaucracies, by applying a neo-Weberian framework based on the ideal type of hyper-rationality (Ritzer and LeMoyne 1991, LeMoyne et al. 1994). A number of authors have highlighted the multifaceted nature of rationalisation processes found in the work of Weber (Kalberg 1980, Levine 1981, Brubaker 1984, Tenbruck 1989). Kalberg (1980) and Levine (1981) identify four distinct types of rationality within Weber's work – practical, formal, substantive and theoretical – which become institutionalised in organisations and give logic to patterns of action[2] (Kalberg 1980). Practical rationality refers to making decisions based on self-interest or expedient grounds. Formal rationality refers to a means-end calculation and is most clearly represented in bureaucratic rules and regulations. Substantive or value rationality refers to actions based on adherence to a certain value system (for example, a commitment to professionalism or equity). Theoretical rationality is based on abstract concepts derived from cognitive processes, such as notions of causal relationships and logical deduction that people use to make sense of their world (Kalberg 1980, Levine 1981).

The ideal type of hyper-rationality refers to a synthesis of the four rationality types to produce a 'whole' that is greater than the sum of its parts[3] (Ritzer and LeMoyne 1991). The term was first used by Ritzer and Walczak (1988) and the concept was significantly elaborated upon by Ritzer and LeMoyne (1991) and LeMoyne et al. (1994). These authors argue that new regimes of management epitomise hyper-rationality by drawing upon and coalescing the four rationality types to produce a hyper-rationalised labour process. The empirical findings of the study presented here are used to critically examine the utility of the concept of hyper-rationality in the context of professional (public) bureaucracies in the health sector.

Methods

The data were collected as part of a government-funded independent evaluation of the BPHS programme. The BPHS programme, which operated between 1993 and 1996, aimed to develop organisational strategies to improve quality, productivity and cost-effectiveness in healthcare delivery. The programme evaluation was conducted during 1995 and 1996 and consisted of three components: case studies of the funded projects (including interviews with participants), quantitative surveys of the health sector, and interviews with unsuccessful applicants (Germov *et al.* 1998). The data reported here are drawn solely from the case study component of the programme evaluation, which essentially consisted of nested case studies of a selection of the funded projects completed at the time the programme evaluation was conducted. Using a standardised case-study protocol, evaluation team members visited the funded projects and collected project documentation (such as minutes of meetings and project reports), and conducted semi-structured interviews with all available project staff[4]. Participation of project staff in the programme evaluation was a requirement of project funding, and the data that form the empirical part of the findings reported here were collected on that basis. A total of 71 people were interviewed, comprising 16 nurses, 12 physiotherapists, 10 pharmacists, eight occupational therapists, five medical specialists, five social workers, four health managers (two hospital CEOs and two general managers), four dietitians, three psychologists, two general practitioners, one Aboriginal health worker and one patient advocate. The majority (60 out of 71) of the participants were female, reflecting the sexual division of labour in the health sector workforce. The number of participants varied between each of the 11 cases because of the different nature and size of the funded projects. All interviews were conducted on the premises of the public sector health organisation where the project was located. Five projects were located in urban teaching hospitals and are referred to as the Well-being (WB), Intra-sectoral (IS), Dialogue (DE), Pathway (PW) and Internal Customer (IC) projects. Three were in community health services, referred to as the Comprehensive Service (CS), Integration (IN) and Technology projects (TY). Two were in rural hospitals, the Counselling (CG) and Community (CY) projects, and one was in an urban mental health service, the Forum project (FM).

Anonymity and confidentiality for study participants were ensured by the use of author-chosen pseudonyms for all projects, and by identifying participants' responses only by occupational designation, sex and project pseudonym in the reporting of results. Interview participants comprised key informants (project leaders) and staff who were directly involved in each funded project. Permission to tape-record interviews was sought from participants, and if this was declined, interview responses were recorded in written field notes. Typed transcripts of each individual interview and related notes were produced at the earliest practicable time and grouped under related interview questions by the researcher who had collected the data. For the purposes of this study, the data were then coded using thematic analysis by the author who re-read the grouped transcripts and coded for themes arising from participants' responses. Direct interview quotes are used to support the narrative throughout this chapter. For the sake of clarity and readability, extraneous comments and the usual quota of ums and errs have been omitted, without altering the meaning of what participants said. While there is considerable debate in the literature over issues of reliability and validity in qualitative research (see Lincoln and Guba 1985, Marshall and Rossman 1995), in this study they were addressed by using a multiple case design, interviewing a number of participants at each site and the collection of supporting documentary material. As a further validity check, participants were provided with a transcript of their interview for comment and clarification. A few participants suggested modifications and their transcripts were subsequently amended. The Human Research Ethics Committee of The University of Newcastle, Australia, approved the study methodology.

Findings

The findings are organised to explore the utility of the hyper-rationality concept by discussing examples of the four types of rationality evidenced in the BPHS projects. The findings show the presence of hyper-rationality, in varying degrees, among all the projects, which is suggestive of an intensification of rationalisation processes in professional (public) bureaucracies. However, I also note the limitations of such a perspective, and the results section concludes with a discussion of some examples of 'irrationality' from the projects.

Practical rationality: opportunism, professional projects and teamwork
One of the novel features of this study was that participants voluntarily chose to introduce the managerial strategy of best practice into their workplace. Therefore, the experience of managerialism reported here was not the outcome of a predetermined and imposed process. The BPHS programme provided short-term funding for the development of 'dissemination projects' that were meant to adapt the principles of best practice to healthcare work, which if they proved successful, could then be implemented more widely throughout the health system. Such a process gave participants a much greater scope to exert their agency and influence on the nature and outcomes of managerialism compared with what is usually reported in the literature.

Participants commonly admitted that their involvement in the best practice projects was opportunistic in terms of securing extra funds for their workplace. Few participants had actually heard of best practice before their involvement in the funded projects. As one participant put it: 'even though no-one knew what it was, we thought we'd give it a shot' (Dietitian, female, IN). The government program was viewed as a 'rare opportunity to get extra funds' (Nurse, female, CG), particularly when 'faced with a declining budget . . . and a projected increase in patient throughput' (Project leader, female, PW), or in the context of 'rumours that we might be contracted-out' (Pharmacist, male, IC). One participant commented:

> We needed to find a smart way to handle funding cuts and make the department viable . . . we didn't want to be the lackeys of bean counters (Occupational therapist, female, TY).

According to Kalberg (1980: 1152), 'a practical rational way of life accepts given realities and calculates the most expedient means of dealing with the difficulties they present'. Participants often conveyed a reluctant acceptance of the prevailing political and economic environment that was responsible for the work pressures they confronted. Therefore, the importance of access to extra funding cannot be underestimated, especially for the majority of participants who were allied health practitioners and nurses in a health system that prioritises the funding of medical therapies. As one participant said, 'If you're not a doctor, it's almost impossible to be successful in getting any money for research, let alone anything else' (Social worker, female, CG).

Participants also viewed their project involvement as an opportunity to 'promote what we do so that our expertise can be fully used' (Nurse, female, PW), expand professional practice 'to act more as consultants rather than technicians' (Pharmacist, male, IC), and as a means of 'giving better patient care' (Nurse, female, FM). Such concerns imply that the BPHS programme was co-opted by some nursing and allied health professionals as part of a wider professional project. Harrison and Ahmed (2000) and Flynn (2002), among others, note how managerialism can be used by health professionals to legitimise professional practice, albeit by agreeing to be subject to stand-ardised performance measures and work protocols.

Most, though not all, of the participants in this study accepted that managerial strategies could be adapted to their professional work, reflecting the negotiated order of professional bureaucracies. Some, however, remained unconvinced of the utility of best practice, as the following comments suggest:

> It's not going to make one iota of difference . . . It's politics that's driving things and I don't think best practice is going to change that (Nurse, female, WB).

> This stuff's just a 'drop in the ocean', 'small bickies' . . . another passing fad (Dietitian, female, DE).

These participants were very sceptical about the concept of best practice, especially the actual phrase 'best practice', with some preferring to use 'better practice' or 'effective practice' in order to avoid the élitist overtones of being 'the best', which they believed undermined the collaborative nature of healthcare work.

Ritzer and LeMoyne (1991) point out that teamwork can tap into employees' practical rationality by acting as a forum to synthesise the experience and skill of individual employees to solve problems based on practical means-end calculations. All 11 projects adopted teamwork structures and it was not uncommon for project leaders to emphasise the benefits of teamwork:

> The simple act of getting together and discussing workplace issues in a regular, systematic way has been of great benefit to us. But the real difference has been the collective form of decision-making that's emerged (Project leader, female, IC).

Some caution needs to be noted here in terms of participant responses, especially those of project leaders, given their vested interest in providing a positive depiction of working with colleagues and project outcomes in general. Nonetheless, participant experiences of teamwork among the projects varied considerably, from positive examples of self-managing teams to dysfunctional teams rife with personality clashes and power struggles. Given the voluntary nature of participants' involvement in the BPHS programme, many appeared willing to offer frank and critical comments. For example, in some projects the experience of teamwork was notional, being 'in name only' [Nurse, female, CS], whereby participants 'followed directions' issued by their project leader [Occupational therapist, female, WB]. Unsurprisingly, best practice was unable to overcome the informal power relations of the workplace, which are likely to persist no matter what managerial strategy is employed (*cf.* Blau 1963, Albrow 1970). An extensive literature has also documented the variability of team structures and the potential for conflict and coercion by team processes, highlighting that teamwork is not a universally positive experience or a panacea for workplace problems (*cf.* Knights and McCabe 1998, Delbridge and Lowe 1997, Barker 1993, Sewell and Wilkinson 1992). While the experience of teamwork by study participants undermines any unitary rationality in respect of work organisation, as implied by Ritzer and LeMoyne, in the majority of projects the collaborative work of team members laid the groundwork for the formal rationalisation of aspects of professional work.

Formal rationality: performance measurement, work protocols and contracts
All projects involved the formal rationalisation of some work tasks through the introduction of various types of performance measurement, work protocols or contractual relations. Two of the projects, the Pathway and Internal Customer projects, will be discussed as they provide representative examples of the way formal rationality was introduced among the projects.

The Pathway project developed what they termed a 'critical pathway' for the physiotherapy treatment of stroke patients. A critical pathway was a set of clinical practice guidelines that mapped the provision of services in a chronological manner in terms of diagnosis and treatment for a specific condition, from acute clinical services to rehabilitation and community support. The pathway also served to identify and remove variance in patient treatment and aimed to achieve optimal patient care. It was a 10-page document and consisted of a mix of validated functional assessment scales (drawn from professional journals), personal patient details, a daily treatment record and discharge planning information. As the project leader stated, 'Performance measures were built into daily tasks so that we could make the recording of information automatic' (female, PW). The high level of detail and the comprehensive range of protocols for patient treatment effectively acted as a series of clinical prompts for staff, as the following comments indicate:

> It's more of a team approach to doing things, with checks and balances built in (Occupational therapist, female, PW).

> It's a form of protection for the patient and for us, by making sure that all the steps that are meant to be followed are (Physiotherapist, female, PW).

While the critical pathway effectively limited the clinical autonomy of individual health professionals, it represented a form of collective autonomy since it was devised by the participants themselves. Ultimately, the pathway served to standardise professional work through the use of protocols and performance measures that were integral to the pathway process.

The Internal Customer project provides a further example of the enhancement of formal rationality found among the projects. It was located in the Pharmacy department of a large teaching hospital and involved the customisation of service delivery to an internal 'customer' (the Paediatrics department), which was formalised by a Service Delivery Agreement (SDA). The SDA outlined the cost and range of services provided, clarified the resource constraints under which the Pharmacy department operated and stipulated service delivery standards and performance indicators. The contractual approach of the SDA had the effect of making the work of pharmacy professionals subject to greater visibility, calculation and control. The SDA listed a range of performance measures, such as guaranteed timeframes for the preparation of intravenous pharmaceutical and nutritional solutions, response times for filling drug orders and the availability of pharmacists for patient education regarding medication use. It also included a range of work protocols for pharmacists to follow, such as a format for patient-specific labelling, dosage instructions, costs per dose, cautionary information regarding drug interaction and side effects, indications for appropriate use and alternative drug therapies available.

There was a genuine belief among the project participants that the work protocols and performance measures in the SDA, despite the standardisation of professional practice they entailed, ultimately reinforced their professional autonomy. For the participants, the 'pay-off' for work standardisation was the potential for work enrichment:

> We wanted to be more than just a drug dispensary so we decided to take our future into our own hands and work together to find a way to thrive professionally. I wanted to create a flowering professional environment and to be appreciated for our expertise. That's the 'pay-off' in the long run (Project leader, female, IC).

As the project leader implies, the SDA was a survival strategy in the context of budget pressures and the possibility of being contracted-out:

> Pharmacy has been a bit of a 'whipping boy'. It's usually seen as a major area of rising hospital costs even though we simply fulfil the requests made of us by others . . . There's been a rumour that Pharmacy might be contracted-out . . . We want to show we are doing all we can to offer quality and cost-effective services (Project leader, female, IC).

Ezzy (1997: 440) notes there is a need to understand the subjective experiences of employees, whereby they may internalise managerial norms and goals, but they may also have their own agendas and 'manipulate and utilise cultural discourses to further their own ends' (1997: 441). The expansion of professional practice, the defence of professional domains and the protection of people's actual jobs featured highly among the reasons participants gave for supporting self-imposed work standardisation. As one participant cogently put it, 'the team approach meant we could still keep our ability to make clinical decisions, but it was done at a team level' (Pharmacist, female, PW). Given that participants were able to specify the nature of the work protocols and performance measurements introduced, the enhancement of formal rationality over their work can be viewed as a form of *collectivised* clinical autonomy (*cf.* Freidson 1994, Dent 1998). Since professions have traditionally claimed the goal of high-quality work, the self-imposition of formal rationality over aspects of their work is not incongruous with professional practice as it can serve to legitimate professional expertise because it requires professionals' co-operation (*cf.* Flynn 2002, Harrison and Ahmad 2000).

Substantive and theoretical rationality: professional values and training
Substantive rationality concerns action 'shaped by a coherent set of social values' (Ritzer and Walczak 1988: 4). It is often reinforced with a theoretical rationality that 'involves a conscious mastery of reality through the construction of increasingly precise abstract concepts' (Kalberg 1980: 1152), such as training in new skills and knowledge, and professional codes of conduct. Specific managerial strategies exhibiting substantive and theoretical rationality were less prominent among the projects than examples of

practical and formal rationality. This can partly be explained by the nature of a professional workforce, which by definition is already highly trained and imbued with ethical and value commitments to their profession.

The most common form of substantive rationality among the projects was the expression of 'groupism', which Ritzer and LeMoyne (1991) define as a value commitment to systematic worker co-operation. In the context of professional work, 'groupism' was evidenced by professional solidarity in the form of peer review, team commitment and the widespread support of collective clinical autonomy. There is an apparent overlap here between the practical rationality of teamwork and the value rationality of groupism. Admittedly, such a categorisation can appear problematic at first, but it represents the nature of hyper-rationality – as a synthesis of rationality types – which in real life is not always as clearly distinguished as an ideal type framework may imply.

Four of the projects exhibited specific forms of substantive rationality, such as commitments to gender equity in the Counselling project on treating sexual assault patients, patient advocacy for mental health residents in the Forum project indigenous community consultation and participation in the Community project and workplace safety in the Technology project. Value rationality was reinforced in these projects by theoretical rationality, through specialised training in areas such as trauma counselling, cross-cultural awareness and occupational health and safety (OHS). Other projects provided generic skills-based training in areas such as conflict resolution, informational technology and financial management.

A specific example of theoretical rationality was evident in the Well-being project, which introduced a comprehensive OHS training package into its workplace. The project developed a new preventative approach to OHS by linking it to performance appraisal and offering financial incentives to organisational units that could demonstrate appropriate processes were being followed. However, some project team members were sceptical of the linkage between OHS and performance appraisal, suggesting that the focus on individual behaviour obscured wider workplace problems:

> It's hard for us to take it all too seriously when jobs have been cut, the machinery's old and there's constant pressure to take short-cuts to get things done because there's a lack of staff or proper machinery (Occupational therapist, female, WB).

Budgetary constraints across the organisation had meant that some occupational hazards could not be addressed without extra resources, which were not forthcoming. The project explicitly avoided wider structural factors such as workload pressures and the impact of staff cuts on OHS (see Quinlan *et al.* 2001). As one member of the project team bluntly put it, 'best practice doesn't come cheap ... to really improve things costs money' (Nurse, female, WB). The experience of the Well-being project highlights the limitations of an ideal type framework, whereby the influence of the wider organisational environment and unintended or irrational factors tend to be marginalised.

The irrationality of hyper-rationality
Among the projects, some attempts to rationalise professional work resulted in unintended and 'irrational' outcomes. Ritzer (1993: 121) argues that 'rational systems inevitably spawn a series of irrationalities that serve to limit, ultimately compromise, and perhaps even undermine, their rationality'. The main 'irrationalities' that participants in this study noted included: work intensification and 'change fatigue'; teamwork problems (as noted earlier), and the marginalisation of qualitative aspects of service delivery due to the reliance on quantitative performance measures.

Most participants made reference to the intensification of work they had experienced in recent years. Somewhat predictably, involvement in the best practice projects had also added to staff workloads and a number 'complained that best practice has made things worse because it's on top of their normal duties' (Project leader, female, FM). Another participant commented:

> Sometimes you just feel like saying 'make one more change and I'll scream'. There's a point where you just have to say let's stop for a while otherwise it becomes a blur. You never get to see if something really works because the next new thing is just around the corner . . . we're suffering 'change fatigue' (Physiotherapist, female, PW).

This statement proved prescient as a change of federal government from Labor to the conservative Liberal-National coalition in 1996 resulted in the abolition of the BPHS programme; a victim of an 'across-the-board' public sector budget cut and the perception that it was closely aligned to Labor party interests because of the support of the programme by public sector unions. While anecdotal evidence indicates some projects had a lasting impact in the specific workplaces that were funded, the lack of continuing funds meant that project findings were not disseminated throughout the health system. Unfortunately, this is not an uncommon occurrence in Australia, where many reform initiatives are funded on a short-term basis (usually to fit into three-year federal electoral cycles). As a number of comments from participants indicated, best practice was one of many 'passing fads' and budgetary constraints meant that 'at the end of the day, despite all your hard work, you're faced with fewer staff and less money' (Occupational Therapist, female, WB). Without sustained programme funding, long-term strategies are difficult to sustain and eventually atrophy as staff turnover displaces organisational memory. This is often exacerbated by the next round of public sector restructuring and reform initiatives that lead to 'change fatigue' and cynicism among staff.

While hyper-rationality involves a coalescence of rationality types, it remains possible that particular types of rationality, such as formal and substantive rationality, conflict with one another as in the case of quantitative and qualitative aspects of healthcare, as the following comment suggests:

> The whole thing's turning upside-down so that what's important is getting the paperwork right so we get a 'beautiful set of numbers'. But what about the service part of it, the personal relationships with patients, the attention, the way we tailor things to individual needs? You can't measure that and so it gets lost (Nurse, female, IN).

As the examples of irrationalities show, hyper-rationality is unlikely readily to overcome conflicting interests or unintended consequences stemming from organisational change. As one participant surmised: 'I fear we're being sucked in by all this best practice stuff that sounds okay on the surface, but I think it feeds right into making us work harder for the same money, or even doing ourselves out of a job' (Nurse, female, CS).

Discussion: towards a hyper-rational professional bureaucracy?

Table 1 summarises the main examples of the four rationality types found among the 11 projects that underpinned the introduction of best practice managerial strategies, yet the question remains whether this is suggestive of an over-arching trend towards hyper-rationality. Practical rationality manifested itself in teamwork and the pragmatic adoption of best practice managerial strategies. Formal rationality figured prominently among the projects

Table 1 *Examples of the four rationality types among the projects*

Rationality types	Definition	Application
Practical	Expediency	Pragmatic adoption of best practice
		Teamwork
Formal	Rule-based	Work protocols
		Performance measurement
		Contractual relations
Substantive	Value-based	'Groupism'
		Gender equity
		Patient advocacy and participation
		Indigenous consultation and cultural awareness
		Workplace safety
Theoretical	Abstract concepts	New models of service delivery
		Generic training: IT and interpersonal skills

Source: Adapted from Kalberg (1980), LeMoyne *et al.* (1994)

and is inherent in various activities, including the use of work protocols, performance measures and contractual relations. Substantive rationality was particularly identified in groupism, which privileged a collective form of professional autonomy over individual autonomy, as well as a small number of project-specific value commitments (gender equity, patient advocacy, cultural awareness and workplace safety). The fourth type of rationality, theoretical rationality, tended to support the other types through generic and specific forms of staff training.

The ideal type of hyper-rationality implies that the co-existence, inter-relation and synthesis of the four rationality types result in an enhanced standardisation of the labour process. The empirical findings, however, indicate the four rationalities existed in varying scope and intensity among the projects, which implies there are degrees of hyper-rationalisation. Ritzer and LeMoyne (1991: 112) maintain that the four rationality types should be present 'to a high degree' for hyper-rationality to exist. However, they do not indicate how this should be determined, and therefore such stipulations are problematic. The findings suggest that the four rationality types were not of equivalent intensity and that their presence was not always conducive to a unilateral rationalisation process (as the 'irrationalities' noted earlier indicated). As could be expected, there was some overlap between rationality types, such as the practical rationality of teamwork and the substantive rationality of groupism, though it can be argued that such examples show the coalescence of rationalities and the mutually reinforcing nature of hyper-rationality components. Nonetheless, while examples of all four rationality types were present in each project, theoretical rationality proved to be the least salient feature, with more than half of the projects making use of only generic training in IT and interpersonal skills. As already noted, this reflects the highly-educated nature of a professional workforce. It also indicates that pre-existing characteristics of individuals and workplaces play a significant role in how managerial strategies are interpreted and implemented. Moreover, the issues of budgetary constraints, work intensification and change fatigue raised by some participants expose the limitations of an ideal type schema; and highlight the central importance of taking account of the impact of wider organisational and political contexts. This is particularly pertinent given that the BPHS programme funded the introduction of organisational change within sub-units of often much larger organisations, that inevitably had pre-existing managerial strategies and workplace cultures.

Ritzer and LeMoyne (1991: 114) pose the question: 'does the world confront an iron cage of hyperrationality that dwarfs in terms of its problematic dimensions Weber's iron cage of formal rationality?' They are ambivalent about the implications of hyper-rationality, viewing it, as did Weber with formal rationality, as having the potential to be both liberating and constraining. For example, a case can be made that the rationalisation of professional work is long overdue, particularly in terms of ensuring quality patient care. Yet given the study findings, it is worth questioning whether use of the term 'hyper-rationality' is appropriate. While there are limitations with this neo-Weberian framework, it does highlight the multiple forms of rationalisation taking place that enhance the standardisation of professional work. It is this acknowledgement of multiple and coalescing rationalities as underpinning the logic of managerialism processes, in professional bureaucracies at least, that represents the novel contribution of the hyper-rationality concept. Even though the scope and intensity of the four rationality types may vary, these multi-rationalities operate with a unitary purpose of work standardisation, and are likely to synthesise and intensify the rationalisation process – which the term hyper-rationality attempts to convey. Given the complex and indeterminate nature of professional work, hyper-rationalisation is only possible with the active participation of health professionals themselves. As one participant aptly put it: 'What we basically did was systemise what many of us were doing anyway' (Physiotherapist, female, PW). Flynn (2002: 169) comes to a similar conclusion in respect of clinical governance, stating it 'is geared towards "modernising" professional self-regulation, so that while clinicians (and the professional bodies and Royal Colleges) will help set the standards, the system will also identify "lapses"'.

In the specific organisational and political context of the projects discussed here, managerial processes underpinned by a logic of hyper-rationalisation are circumscribing the clinical autonomy of individual health professionals, while a collective form of professional autonomy remains intact; a finding echoed by other studies from a range of theoretical perspectives (*cf.* Flynn 2002, Harrison and Ahmad 2000, Dean 1999, Dent 1998, Freidson 1994).

Conclusion: re-negotiating order in professional bureaucracies

The study provided a valuable opportunity to examine organisational change in professional (public) bureaucracies, drawing on a diverse range of sites and participants. Whereas most studies of managerialism have focused on medicine and nursing, this study shows that the same processes are impacting on allied health professions as well. The findings add to a growing literature that indicates the last bastion of health professionals' autonomy, clinical practice, is increasingly subject to constraint. What is particularly of interest is that the standardisation of professional practice is partly self-imposed, though this can be understood as a 'survival strategy' within the wider context of government and organisational policies that have led to budgetary and staffing reductions. Health professionals' adoption of managerial strategies can also be viewed as a pre-emptive attempt at legitimating professional practice by formalising notions of peer review and self-regulation. Managerial processes such as performance measurement are increasingly becoming part of the daily work of many health professionals, but as Dent (1998: 205) notes in reference to doctors, they are able to 'renegotiate their medical autonomy'. While doctors have more scope to negotiate, my study shows that nursing and allied health professionals are similarly able to exercise their agency and produce a 're-negotiated order' that incorporated managerial strategies. As one project leader stated, 'you've got to play the game and try and turn it to your advantage' (female, IC).

The study findings must be considered within the context of certain limitations. The sample of cases was not representative of all health organisations adopting best practice strategies, but was restricted to the funded projects. Furthermore, the case studies only present a 'snapshot' in time of an on-going change process, and thus suffer the constraints of cross-sectional data. The pilot nature of the projects meant that the case studies focused on organisational units restructuring their immediate workplace, rather than on the wider health organisation. The findings are, therefore, context-specific and must be interpreted taking account of the specific internal organisational influences and constraints on the projects studied. As the findings of this study make clear, an adequate understanding of the impact of workplace change is predicated on placing it within a wider organisational and political context, as well as exploring the subjective experiences of organisational actors.

The concept of hyper-rationality presents a novel framework for understanding the underlying logic of managerialism in the context of professional (public) bureaucracies in the health sector. A new hyper-rational iron cage may well be in the making as the processes of managerialism continue. The traditional Weberian bureaucracy may be passé, but the processes of bureaucratisation continue – the iron cage lives on in new bureaucratic forms.

Acknowledgements

This is a revised and expanded version of a paper that was originally presented at the International Sociological Association XV World Congress, Brisbane, Australia, July 7–13, 2002 (Germov 2002). My thanks to Lauren Williams for her incisive comments on drafts of this chapter, to Lois Bryson for her insightful influence on my work, to the editors and SHI anonymous reviewers for their constructive feedback, and to my BPHS program evaluation colleagues and the participants who gave so freely of their time. I take sole responsibility for the views expressed in this chapter.

Notes

1   In Weber's usage, an ideal type refers to the abstract or pure features of any social phenomenon. Weber's classic formulation of the ideal type bureaucracy emphasised its pyramidal structure and role specialisation within a hierarchical division of labour bounded by formal rules and regulations (see Weber 1968/1921: 221–3).

2 These are Kalberg's (1980) terms and I use them for the sake of clarity. Levine (1981) adopts slightly different terminology, using conceptual and instrumental rationality in place of theoretical and practical rationality, though he defines them in an almost identical way.

3 Unlike LeMoyne *et al.* (1994) and Ritzer and Walczak (1989), I prefer to hyphenate the term 'hyper-rationality' to denote clearly its neo-Weberian lineage.

4 As one of the consultants employed to conduct the case studies, I participated in designing the case study protocol and in collecting the data, and was one of the co-authors of the programme evaluation report (see Germov *et al.* 1998).

References

Allsop, J. and Saks, M. (eds) (2003) *Regulating the Health Professions*. London: Sage.

Albrow, M. (1970) *Bureaucracy*. London: Pall Mall Press.

Allan, C. (1998) The elasticity of endurance: work intensification and workplace flexibility in the Queensland public hospital system, *New Zealand Journal of Industrial Relations*, 23, 3, 131–51.

APS, Australian Public Service Commission (2003) *The Australian Experience of Public Sector Reform*. Canberra: Australian Public Service Commission.

Barker, J.R. (1993) Tightening the iron cage: concertive control in self-managing teams, *Administrative Science Quarterly*, 38, 3, 408–37.

Beattie, B. (2000) What impact has managerialism had on a New South Wales Area Health Service? *Australian Health Review*, 23, 4, 170–5.

Belcher, H. (2005) Power, politics, and health care. In Germov, J. (ed.) *Second Opinion: an Introduction to Health Sociology*, 3rd Edition. Melbourne: Oxford University Press.

Blau, P.M. (1963) *The Dynamics of Bureaucracy*, 2nd Edition. Chicago: University of Chicago Press.

Braithwaite, J. (1997) *Workplace Industrial Relations in the Australian Hospital Sector*. Sydney: School of Health Services Management, University of New South Wales.

Braithwaite, J. and Hindle, D. (1998) Casemix funding in Australia: time for a rethink, *Medical Journal of Australia*, 168, 558–60.

Brubaker, R. (1984) *The Limits of Rationality: an Essay on the Social and Moral Thought of Max Weber*. London: George Allen and Unwin.

Bryson, L. (1987) A new iron cage? Experiences of managerial reform, *Flinders Studies in Policy and Administration*, 3, March, 17–45.

Camp, R.C. (1989) *Benchmarking: the Search for Industry Best Practices that lead to Superior Performance*. Milwaukee: ASQC Quality Press.

Coburn, D., Rappolt, S. and Bourgeault, I. (1997) Decline vs. retention of medical power through restratification: an examination of the Ontario case, *Sociology of Health and Illness*, 19, 1, 1–22.

Considine, M. (1990) Managerialism strikes out, *Australian Journal of Public Administration*, 49, 2, 166–78.

Considine, M. and Painter, M. (eds) (1997) *Managerialism: the Great Debate*. Carlton South: Melbourne University Press.

Courpasson, D. (2000) Managerial strategies of domination: power in soft bureaucracies, *Organization Studies*, 22, 141–61.

Dean, M. (1999) *Governmentality: Power and Rule in Modern Society*. London: Sage.

Delbridge, R. and Lowe, J. (1997) Manufacturing control: supervisory surveillance on the 'new' shopfloor, *Sociology*, 31, 3, 409–26.

Dent, M. (1998) Hospitals and new ways of organisation of medical work in Europe: standardisation of medicine in the public sector and the future of medical autonomy. In Thompson, P. and Warhurst, C. (eds) *Workplaces of the Future*. London: Macmillan.

Dertouzos, M., Lester, R. and Solow, R. (1989) *Made in America: Regaining the Productive Edge*. Cambridge, Mass.: MIT Press.

DHSH, Department of Human Services and Health (no date) *Best Practice in the Health Sector Information Kit*. Canberra: Department of Human Services and Health.

DIR/AMC, Department of Industrial Relations and the Australian Manufacturing Council (1992) *International Best Practice: Report of the Overseas Study Mission*. Canberra: Department of Industrial Relations.

Draper, M. (1992) *Casemix, Quality and Consumers*. Melbourne: Health Issues Centre.

Exworthy, M. and Halford, S. (eds) (1999) *Professionals and the New Managerialism in the Public Sector*. Buckingham: Open University Press.

Ezzy, D. (1997) Subjectivity and the labour process: conceptualising 'good work', *Sociology*, 31, 3, 427–44.

Ferlie, E., Pettigrew, A., Ashburner, L. and Fitzgerald, L. (1996) *The New Public Management in Action*. Oxford: Oxford University Press.

Flynn, R. (2002) Clinical governance and governmentality, *Health, Risk and Society*, 4, 2, 155–73.

Freidson, E. (1994) *Professionalism Reborn: Theory, Prophecy and Policy*. Cambridge: Polity Press.

Germov, J. (2002) Second wave managerialism and the hyper-rationalisation of profession work: case studies from the Australian public health sector, *Sociological Abstracts*, Conference paper presented at the International Sociological Association XV World Congress, Brisbane, Australia, 7–13 July.

Germov, J., Heiler, K. and Pickersgill, R. (1998) *The Evaluation of the Best Practice in Health Program*. Working Paper 54, Sydney: Australian Centre for Industrial Relations, Research and Training, University of Sydney.

Halligan, J. (1997) New public sector models: reform in Australia and New Zealand. In Lane, J. (ed.) *Public Sector Reform: Rationale, Trends and Problems*. London: Sage.

Harrison, S. and Ahmad, W. I. (2000) Medical autonomy and the UK state, 1975 to 2025, *Sociology*, 34, 1, 129–46.

Harrison, S. and Dowswell, G. (2002) Autonomy and bureaucratic accountability in primary care: What English general practitioners say, *Sociology of Health and Illness*, 24, 2, 208–26.

Hoggett, P. (1996) New modes of control in the public service, *Public Administration: An International Quarterly*, 74, Spring, 9–32.

Hughes, O. (2003) *Public Management and Administration: an Introduction*, 3rd Edition. New York: Palgrave.

Kalberg, S. (1980) Max Weber's types of rationality: cornerstones for the analysis of rationalization processes in history, *American Journal of Sociology*, 85, 5, 1145–79.

Keating, M. (1990) Managing for results in the public interest, *Australian Journal of Public Administration*, 49, 4, 387–98.

Knights, D. and McCabe, D. (1998) Dreams and designs on strategy: a critical analysis of TQM and management control, *Work, Employment and Society*, 12, 3, 433–56.

LeMoyne, T., Falk, W.W. and Neustadtl, A. (1994) Hyperrationality: historical antecedents and contemporary outcomes within Japanese manufacturing, *Sociological Spectrum*, 14, 3, 221–40.

Levine, D. (1981) Rationality and freedom: Weber and beyond, *Sociological Inquiry*, 51, 5–25.

Light, D.W. (2000) The medical profession and organizational change: From professional dominance to countervailing power. In Bird, C., Conrad, P. and Fremont, A.M. (eds) *Handbook of Medical Sociology*, 5th Edition. New Jersey: Prentice Hall.

Lin, V. and Duckett, S. (1997) Structural interests and organisational dimensions of system reform. In Gardner, H. (ed.) *Health Policy in Australia*. Melbourne: Oxford University Press.

Lincoln, Y. and Guba, E. (1985) *Naturalistic Enquiry*. Beverley Hills: Sage.

Marshall, C. and Rossman, G.B. (1995) *Designing Qualitative Research*, 2nd Edition. Thousand Oaks Park: Sage.

Mintzberg, H. (1979) *The Structuring of Organizations*. Englewood Cliffs, New Jersey: Prentice Hall.

O'Brien, J. and Fairbrother, P. (2000) A changing public sector: developments at the Commonwealth level, *Australian Journal of Public Administration*, 59, 4, 59–66.

O'Brien, J. and O'Donnell, M. (2000) Creating a new moral order? Cultural change in the Australian Public Service, *Labour and Industry*, 10, 3, 57–76.

O'Brien, J. and O'Donnell, M. (2002) Towards a new public unitarism: employment and industrial relations in the Australian Public Service, *Economic and Labour Relations Review*, 13, 1, 60–87.

O'Donnell, M., Allan, C. and Peetz, D. (2001) New Public Management and workplace change, *Economic and Labour Relations Review*, 12, 1, 85–103.

O'Donnell, M. and Shields, J. (2002) Performance management and the psychological contract in the Australian federal public sector, *Journal of Industrial Relations*, 44, 3, 435–57.

Osborne, D. and Gaebler, T. (1992) *Reinventing Government: How the Entrepreneurial Spirit is Transforming the Public Sector from Schoolhouse to State House, City Hall to Pentagon*. Reading: Addison Wesley.

Paterson, J. (1988) A managerialist strikes back, *Australian Journal of Public Administration*, XLVII, 4, 287–95.

Peters, T.J. and Waterman, R.H. (1982) *In Search of Excellence: Lessons from America's Best-run Companies*. New York: Harper and Row.

Pollitt, C. and Bouckaert, G. (2000) *Public Management Reform: a Comparative Analysis*. Oxford: Oxford University Press.

Quinlan, M., Mayhew, C. and Bohle, P. (2001) The global expansion of precarious employment, work disorganization, and consequences for occupational health: a review of recent research, *International Journal of Health Services*, 31, 2, 335–414.

Ritzer, G. (1993) *The McDonaldization of Society*. California: Pine Forge Press.

Ritzer, G. and Walczak, D. (1988) Rationalization and the deprofessionalization of physicians, *Social Forces*, 67, 1–22.

Ritzer, G. and LeMoyne, T. (1991) Hyperrationality: an extension of Weberian and neo-Weberian theory. In Ritzer, G. (ed.) *Metatheorizing in Sociology*. Massachusetts: Lexington Books.

Rose, N. and Miller, P. (1992) Political power beyond the state: problematics of government, *British Journal of Sociology*, 43, 2, 173–205.

Sewell, G. and Wilkinson, B. (1992) Someone to watch over me: surveillance, discipline and the just-in-time labour process, *Sociology*, 26, 2, 271–89.

Sheaff, R., Rogers, A., Pickard, S., Marshall, M., Campbell, S., Sibbald, B., Halliwell, S. and Roland, M. (2003) A subtle governance: 'soft' medical leadership in English primary care, *Sociology of Health and Illness*, 25, 5, 408–28.

Stanton, P. (2002) Workplace reform in the public health care sector. In Gardner, H. and Barraclough, S. (eds) *Health Policy in Australia*, 2nd Edition. Melbourne: Oxford University Press.

Strauss, A., Schatzman, L., Ehrlich, D., Bucher, R. and Sabshin, M. (1963) The hospital and its negotiated order. In Freidson, E. (ed.) *The Hospital in Modern Society*. New York: Free Press.

Taylor, F.W. (1947/1911) *Scientific Management*. New York: Harper and Brothers.

Tenbruck, F.H. (1989) The problem of thematic unity in the works of Max Weber. In Tribe, K. (ed.) *Reading Weber*. London: Routledge.

Turner, B.S. (1995) *Medical Power and Social Knowledge*, 2nd Edition. London: Sage.

Walby, S. and Greenwell, J. with Mackay, L. and Soothill, K. (1994) *Medicine and Nursing: Professions in a Changing Health Service*. London: Sage.

Weber, M. (1968/1921) *Economy and Society: an Outline of Interpretive Sociology*. 3 volumes, Roth, G. and Wittich, C. (eds), New York: Bedminster Press.

Weekes, K., Peterson, C. and Stanton, P. (2001) Stress and the workplace: the medical scientists' experience, *Labour and Industry*, 11, 3, 95–120.

Yeatman, A. (1994) The reform of public management: an overview, *Australian Journal of Public Administration*, 53, 3, 287–95.

Chapter 5

What's in a care pathway? Towards a cultural cartography of the new NHS

Ruth Pinder, Roland Petchey, Sara Shaw and Yvonne Carter

[Care pathways] are the geographic maps of managed care. [They] eliminate the boundaries of time and space by removing the walls within a health care setting. [They] rattle our needs for 'turf' and revise our ideas of territory. [They] help us chart the way for truly patient-centred care (Etheridge 1986: 1–3).

There is and can be no such thing as a purely objective map, one that reproduces a pre-existing reality. Choices always have to be made about what to represent and how, and what to leave out. To be included on the map is to be granted the status of reality or importance. To be left off is to be denied (King 1996: 18).

An Emperor wishes to have a perfectly accurate map of the empire made. The project leads the country to ruins – the entire population devotes all its energy to cartography (Borges 1984: 325).

Introduction

A recent exhibition at the British Library, seductively entitled 'The Lie of the Land', illustrated how the appeal of maps lies in the way they hover between fact and fiction. They are not mirrors of nature. Whilst ostensibly timeless, universal and objective, they may better be seen as powerful rhetorical tools imposing their own linear order on to spaces and events. Rather than representing complex behaviours and relationships, they present only inert fixtures from which agency and intentionality have disappeared, history has been erased. De Certeau (1984) puts the matter beautifully: 'Itself visible /the map/has the effect of making invisible the operation that made it possible . . . /In transforming/action into legibility, it causes a way of being in the world to be forgotten' (1984: 121).

What maps silence is just as interesting as what they make visible, for this says a great deal about the compilers of maps, and the difficulties of penetrating the unseen structures which shape people's lives (Wood 1992, King 1996, Black 1997). In cautioning us against mistaking the map for the territory, theorists have powerfully argued that whilst territory may be authoritatively mapped, it can never be truthfully mapped (Harvey 1992).

But more is at stake. As ideological constructs, maps are laced with interests, with at least an eye on re-distributing resources and advancing sectional interests, licensing certain forms of intervention and marginalising (or excluding) others. The questions that maps pose are thus moral as well as epistemological. If maps reflect and sustain power relationships, how might an understanding of mapping and map use as processes of knowledge construction (as opposed to knowledge transfer) enable them to be used positively to promote social justice?

It is precisely these tensions that form the subject matter of this chapter. We approach them dialectically through an analysis of the interplay between maps as quasi-objective tools or faits accomplis, the practice of map-making (the compiling of maps), and the process of reflecting on that practice, that we term 'mapping' (Ingold 2000). This interplay offers a window on the way that maps produce their reality effect 'as long as they are confirmed in their existence by means of recursive acts of belief and practical investment' (Pels 2002: 69).

Health policy, like many other areas of contemporary governance, has not been immune to this proliferation of map-making (Abbott 1988). Techniques of gene mapping, the development of protocols, decision-tree analyses, G.I.S. (geographical information systems), diagnostic and treatment algorithms, and process modelling can all be seen as instances of the mapping process in healthcare, which have as much to do with administering healthcare as understanding the social complexities to which administration gives rise. This chapter is concerned, however, with a specific instance of map-making in health policy, namely the burgeoning use of care pathways[1]. From almost nowhere, apparently, they appear to have become the tool of choice for ensuring (or so it is claimed) quality of care, equity of treatment, optimal allocation of resources and a rational division of labour between healthcare professionals. Simultaneously, it is claimed that they respond even-handedly to concerns for patient safety, variable healthcare quality and spiralling health costs (Bragato and Jacobs 2003).

Care pathways embody an approach to organisations, whose intellectual origins can be traced back at least to the Enlightenment's social engineering model of society with its twin beliefs in constant improvement and rationality. This organisational engineering tradition was manifest in the classical management theory of the late 19th century (Fayol 1949) and in scientific management in the 1920's (Taylor 1911), surfacing again in the form of business process re-engineering in the 1990s (Hammer and Champy 1993). (See McNulty and Ferlie (2002) for a critical case study of BPR in a UK hospital.) In many respects, care pathways are direct descendants of the time-task matrix approach of Gantt charts, developed by Henry Gantt (one of Taylor's assistants) in the early 1890s (Bragato and Jacobs 2003). They emerged in the 1980s in the USA in the context of the development of managed care, as a management tool for rationalising resource utilisation and meeting length of stay parameters imposed by third-party healthcare funders. Originally (1985) 'timelines' (Etheridge 1986), they evolved into 'critical paths' (Woldrum 1987), and by 1990, had been transformed into the 'CareMaps™ System' (Zander 1991). Along the way, what had started out as a tool for planning just nursing care had grown to become 'the core of the *entire* patient care-giving process and its documentation *by all* disciplines (our emphases)' (Zander 1991). At the same time (in an apposite reminder of the closeness of the link between map-making and territorial expansionism), the organisation promoting them itself metamorphosed from The Center for Nursing Case Management into The Center for Case Management.

In the UK, interest in care pathways had begun to emerge piece-meal in the mid-1990s, but the real impetus to their development came with the publication of the White Paper, *The New NHS: Modern and Dependable* (Department of Health 1997). Although it did not refer specifically to care pathways, its aims (national standards, local flexibility, efficiency, quality, patient centred-ness) are strikingly similar to the claims made for pathways (Norris and Briggs 1999).

Website hits have to be viewed as enigmatic artefacts rather than reliable indicators of adoption and use. Nonetheless, they suggest a growing interest in care pathways (Table 1). By 2004, search of the National Electronic Library for Health (NHS Information Authority 2004) revealed records of 237 pathways, spanning the alphabet from abscess via mental health to wrist surgery. Interest in them has been further fuelled by a number of other recent policy developments. These include the drive to standardise professional performance in an era troubled by revelations of professional misconduct post-Shipman, and the shift from professional discretion and variability to a more rules-based, audited practice (Power 1997, Strathern 2000). Further factors in their promotion have been the thrust towards joined-up services (Department of Health 2000) and the constant remaking of professional boundaries, in pursuit of workforce flexibility and professional de-regulation (Gellner and Hirsch 2001). They have also been advocated as a means of making service provision more patient-centred. However, despite their growing popularity, consideration of care pathways to date has been limited, even in the clinical literature, where they have been treated uncritically as

Table 1 *The rise of the care pathway – 'hits' by years*

Year	Number
1997	0
1998	8
1999	78
2000	97
2001	167
2002	377

Source: Department of Health website (www.doh.gov.uk), searched 17 January 2003.

helpful (and technically neutral) tools rather than embodied practices for routing patients through the system (Campbell *et al.* 1998, Owen 1998, Parker 1999). The only reservations that have been voiced about them concern the lack of robust evidence of their effectiveness (Campbell *et al.* 1998).

Theoretically, this chapter draws on a variety of sources, such as Berg's analysis of protocols and medical standardisation (Berg 1997, Berg *et al.* 2000) and Bowker and Leigh Starr's (2000) work on expert classification systems, that view them as embodied practices rather than concrete 'things'. In it we explore how pathways are being envisaged, worked and transformed in the new NHS, and their implications for the way patient and professional subjectivities are currently being fashioned. First, we examine how care pathways de-couple the patient from his/her context, highlighting some aspects of the patient experience, while silencing others. Second, we consider the role of pathways in re-defining inter-professional relationships, especially around the emergent 'hybrid professionals', such as GP specialists, specialist nurses, extended-scope physiotherapists and others. Third, we draw out some of the wider implications (both positive and negative) of pathways as part of the simplifying and disciplinary activity of the state, before offering some cautionary comments about the over-rationalist and sometimes evangelical impetus behind the pathway movement. While not arguing against rational planning per se, we suggest a more modest approach to the possibilities of incorporating pathways into the design of patient care. In particular, we suggest that a critical and processual understanding of pathways – a process cartography (Rundstrom 1993) – might contribute to a more informed appreciation of their potential (and their limitations) as mechanisms for healthcare policy implementation.

Mapping the field: methods, setting and subjects

This study was part of a wider investigation of the referral strategies of the recently created primary care organisations in the 'new NHS'. Its setting was three Primary Care Trusts in the Greater London area, which had been identified (through briefings by regional administrators) as innovators in this respect. One ('Brown') was suburban with a large rural hinterland, one ('Fleming') was semi-rural and one metropolitan ('Jameson'). All three were involved in the process of developing care pathways or else had their development under active consideration. They thus constituted a purposive selection of 'leading edge' primary care organisations in this respect. Interviews were conducted over a period of five months with 25 health professionals and administrators. Three were consultants, eight GPs (including four GP clinical governance and commissioning leads and one GP specialist), one an optometrist, one a public health doctor, one a nurse manager and 11 were PCT Board members. Where informed consent was given, interviews were tape-recorded and transcribed verbatim. In the three instances when consent was withheld, contemporary shorthand notes were taken, which were subsequently transcribed in full. Observations made during site visits and interviews were recorded in shorthand and written up in a fieldwork diary. Interviews and fieldwork notes were written up into interim case study vignettes, which were returned to interviewees for their comments.

Qualitative work in the health services has been appropriately critiqued for the banality of many of its findings (Lambert and McKevitt 2002). The highly stylised, performative nature of the single interview, for example, offers little scope on its own for exploring the tacit meanings so rarely explicated between people, or the way that meanings are always in process of becoming something else. Meaning does not lie on the tip of the tongue. Anthropological studies of organisations have identified profound discrepancies between how people conceive of their organisation and their actual practice, and the inability of rationalist frameworks to capture the subtle, imaginative and symbolic ways people have found to deal with the inconsistencies and contradictions that arise (Ouroussoff 2001: 37).

To deepen our understanding and explore the inevitable slippages between what people said they thought and did and what they 'actually' thought and did, we observed the clinics of an extended scope physiotherapist (ESP), an ophthalmic nurse practitioner, and a specialist dermatology nurse in Brown. We also observed a PCT Board meeting there. In addition to formal observation, we took opportunistic advantage of the occasions for informal observation that arose during fieldwork. The demands of the larger study meant that we were able to investigate only the initial stages of a number of pathways in process, or under active consideration by the PCTs concerned. We were not able to see the process through to official 'completion' or to analyse the finished product. Nevertheless, we did have the opportunity to ask the key health professionals involved in the process to draw their maps, to reflect on them and to articulate some of the assumptions that were built into them. In so doing, we were able to explore the dialectic between maps, map-making and mapping referred to above. We also acknowledge that our representations (Figure 1) are, of necessity, cleaned up and re-ordered versions of what was drawn roughly at interview. Behind the apparent certainties of our lines and arrows and boxes, there lay layers of meaning that were highly conditional, contingent and uncertain.

The study also raised questions about the scope for exploratory research which generates hypotheses rather than merely seeking to confirm pre-existing criteria (Gellner and Hirsch 2001, Agar 1996, Wolcott 1999). As always, gaining access to policy makers is necessarily a sensitive issue. Doors simultaneously opened and closed, and the lead researcher found her contacts carefully orchestrated. In Brown County, for example, a GP specialist was brought in from his annual leave; whereas no amount of negotiation allowed entrée to the ENT consultant at Fleming.

Finally, a perception amongst some participants of research as a practical management tool was at odds with lower key attempts to unravel process.

The Director of Services at Brown County wanted the study to identify 'five things we're doing well and five challenges', whilst the Chief Executive of Fleming County enquired about the publication of results only three months after completion of interviewing there. Difficult questions are raised too about the anticipated use of research to raise organisational profiles. What is the acceptable face of any criticism? Increasingly, qualitative research risks being co-opted not to sensitively illuminate practice, but to proffer simple solutions to complex problems, possibly reproducing the very power structures it needs to challenge in doing so. Weaving a pathway (sic) between these imperatives is the stuff of contemporary health services research.

Exploring the pathway

The pathway and the patient

As with all protocols, it is tempting to view the rationalising and atomising practices of the care pathway as objective and value free (Berg 1997). When, however, managers and practitioners discussed their maps, they revealed a trail of unspoken assumptions, which betrayed their partiality. Figure 1 presents six competing versions of a care pathway (for patients with cataracts), that were elicited or volunteered by different stakeholders in the three healthcare settings.

What is immediately striking about these pathways is how different they are, particularly in respect of which profession occupies prime position (which we discuss below). Despite these differences, there are also important underlying similarities. The first is their tendency towards simplification. This is clearest in maps (i) and (ii) which present the illusion (at least) of closure. They depict a simple, idealised pathway for patients with cataracts, with single start- and end-points, without turnings or crossroads, or choice of routes. All of the pathways also abstract the patient and reify the condition. Predicated on a biomedical and essentialist model of the patient, they all define their situation by their impairment, simplifying the complexity of their lives and reducing their treatment to the clinical disposal of discrete bodily parts. Finally, they assume a large-batch, high-turnover model of healthcare, as the Director of Services at Brown explained:

> It's an easily diagnosable condition, there's lots of them around and it's quite straightforward. There's a large cohort of people all needing the same thing, they need to be funnelled through a process . . . with cataracts what you've got, it's something has happened to their eye. It's like changing the exhaust on your car. When you've had your exhaust changed, you're absolutely fine. Episode over.

He acknowledged that this model of care might not have been feasible for chronic conditions where variability in impairment or co-morbidities might

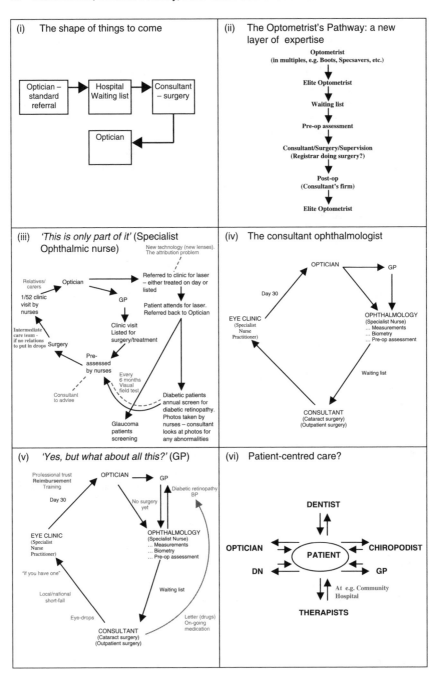

Figure 1 *Competing pathways: the cataract*

complicate the picture. However, by their nature, cataracts lent themselves to the development of a care pathway:

> With a chronic disease it's more difficult. That would be a problem with a pathway. There's a lot of people in those circumstances who might be very different. Whereas a patient with a cataract, a cataract is a cataract is a cataract. You can quite nicely develop a pathway for that.

Yet even with cataracts, the consultant ophthalmologist at Brown acknowledged that patients in different circumstances would need different treatment:

> There are going to be different visual requirements between those who are at home and no longer do any night driving or sewing – that's one thing that gets missed to a certain extent here.

Although it was recognised that variations in age, ethnicity and social class would profoundly affect how patients would experience even a comparatively straightforward condition such as cataracts, these features were admitted into the landscape of care only as after-thoughts. Missing from the maps was any appreciation of how these factors might affect not just the patient experience, but the types and quantities of care required. Also absent was any recognition of the vital role played by informal carers. Ingold's (2000) point about the activity of mapping is nicely illustrated here. The more the maps aimed to furnish a precise, systematic representation of reality, the less true to life they became. Paradoxically it was knowing the detail, not eliminating it, that allowed the practitioner to find his/her way around, and avoid the pitfalls.

Observation and interviews revealed that the one-stop pathway for cataracts relied totally on patients' friends and relatives to administer the post-operative eye-drops. Elderly and disabled patients living on their own, for example, would require referral to the district nurse and intermediate care team, as a GP at Jameson acknowledged (Map v):

> The problem is, even if this works perfectly, and yes, the patient does need a cataract surgery, they have one stop at the clinic here, go on the list and have the operation, things go wrong here. Like the patients don't get their eye drops because the hospital pharmacy's closed the day they're sent home, so they're down here the next day saying, 'They want me to have some eye drops', or, 'Oh, I dropped the bottle', or 'They've run out', or 'I need a prescription', or 'I've got pain and they haven't given me any painkillers', or 'I can't put the drops in because I'm blind or elderly or disabled and who's going to do it?' And so the District Nurse needs to go in. These are the sorts of things that can go wrong . . .

Once these features began to be admitted to the landscape, the care pathway became more complicated. What had started out as a one-size-fits-all model of professionally-delivered care turned out to be dependent for its success upon a complex (and unmapped) network of relatives, friends and informal carers in the community. As soon as the contingencies of patients' lives were overlaid onto the hard rectangles and straight lines of the pathway, the illusion of authoritativeness began to crumble, prompting an exasperated 'It's quite complex, it's supposed to be simple' from the opthalmologist who had been key in promoting the cataract pathway at Brown. The contingent, arbitrary and mechanistic nature of the map was exposed, as the following comment indicates:

> Well I mean it's almost robotic. Ideally I think they would just like to send the eye off and operate on the cataract and then send it back to the patient . . . The whole point about pathways is that they adopt a reductionist view, that you can reduce things just to the cataract and remove them from the rest of the patient's life. I know that patients like to know what's happening, they like to be talked to about it, and they like to understand it and assimilate it, and that doesn't often happen in a pathway (Commissioning lead GP, Fleming).

Nominally, pathways located patients at their centre, treating them as active participants in the care process rather than its passive objects. Certainly, health practitioners appeared to feel that they were expected to demonstrate the patient-centredness of their proposals. Thus, in the preamble to his pathway the consultant opthalmologist commented, 'Essentially I'll do it from the patient's point of view'. Yet only one pathway indicated how matters might look if the patient were placed at its centre (Map (vi)) and even here, patient empowerment turned out to be illusory.

> It's interesting because the whole aspect people forget about in demand management is the patient aspect. I'd put the patient in the middle and then put all the people *acting on him* (our emphasis) (GP Special Projects Lead, Jameson).

We found little evidence that patients had either been involved or consulted in the process of pathway development. In respect of pathways, as in the NHS in general, as the Chief Executive in Fleming acknowledged, patient involvement or consultation was 'tokenism'. An agenda item at a conference on pathways at Brown read, 'Results of patient survey – if time permits'. (Time did not permit, with the result that patients' views were not considered.) The Jameson nurse educator concluded,

> Care pathways aren't integrated with the person, but with the provision of care.

and the Brown optometrist conceded that vested professional interests limited the extent to which pathways could ever genuinely be patient-centred:

They can't be patient-centred. Not really. If they were patient-centred, the consultant would say (*ironically*), 'Go bring them forth, and I will see to them'. The consultant has this great big pyramid below him which keeps him where he is. And no one wants to push him off, because a good consultant is worth his weight in gold.

In all of the maps (but perhaps most clearly in the linear pathways of maps (i) and (ii)) a certain sleight of hand was discernible. By appearing to be cleansed of interests and value-free, the different versions of the pathways gave the impression that all interests represented in them (professionals, patients, employing institutions) were identical and could be simultaneously served. Thus, practitioners enthused about pathways as 'efficient for all concerned' (Chair, PCT, Fleming), or 'No one's the loser: it's a win-win situation all round' (Dermatology Consultant, Brown). However, once these assumptions began to be probed, questions arose:

Doctor: The argument is that you're making it easier for the patient, you're wasting less time, but it really often is just to make the service more efficient so that you see more patients.
RPi: So who are these pathways for?
Doctor: Well I think they are to try and speed things up for patients, but they really are to help the institution run more efficiently I think. They give the impression that they're very straightforward.

The tension between patient-centredness and the organisational imperative of efficiency was also identified by the public health consultant at Fleming:

I don't think the lay person has an idea of what's being done . . . the patient perspective is being overlooked. The question is, is the pathway to manage health service resources better or is it to support the patient better? If there isn't a patient-centred perspective then we could lose the whole point of what patient pathways are for.

Opening up new worlds: territoriality and professional boundaries in the New NHS

Overlying these similarities are some important differences between the pathways, most obviously in the professionals who are included or excluded from them. For instance, maps (i) and (ii) exclude GPs altogether, while maps (iv) and (v) relegate them to the margins of a world that is ophthalmology-centric. Maps (ii) and (iii) introduce new layers of expertise into the traditional mix of services (the 'elite optometrist' and the 'specialist ophthalmic nurse', respectively), who, not surprisingly perhaps, turn out to play central roles in the care pathways being proposed.

The emergence of hybrid professionals, such as nurse specialists, GP specialists and extended scope physiotherapists, who are increasingly taking on tasks traditionally performed by doctors (or specialists), is one of the key features of contemporary healthcare (Leicht and Fennell 2001). They thus represent a challenge not only to the existing division of clinical labour, but to established hierarchies of prestige and power, and current distributions of reward and resources. In contradiction to the claim that they eliminate professional turf wars (Etheridge 1986), it was perhaps inevitable that pathways would generate opportunities for a remapping of professional boundaries.

Among the emergent hybrid professionals a general sense of excitement was in the air. Reconfiguration of services offered chances to spread their wings, and to explore new ways of working. Part of the attraction was the opportunity for personal development, as the specialist ophthalmic nurse at Brown made clear:

> I just think it's nice for the nurse to be able to do more than the basic bedside things. If you've got a special interest in something, it's nice to take that forward, specially in the things that I've achieved. If I'd have stayed a normal nurse, there's no way that I'd have been able to do the things that I do.

Expanding one's professional role in this way, however, was not a selfish matter of personal (or even professional) aggrandisement. It also helped *the whole system* to function more efficiently, as the GP specialist from Fleming explained:

> One consultant, when he looked back, wondered how he coped without us because his local clinic has got a waiting time of over 60 weeks. He traces the amount of work I've done for him in the last couple of years, and he's actually quite pleased.

Thus, pathways furthered the sectional interests of their map-makers without appearing to do so, enabling them to maintain the appearance of neutrality while annexing professional territory (or seeking to do so). We have already noted this feature of pathways in our discussion of the cataract pathway. It emerges even more clearly in the case of the extended scope physiotherapist (ESP) at Brown. Hovering between reality and representation, the map she drew and re-drew, as we talked, placed her centre stage, whilst seemingly off-stage, creating an aura of truth and precision, and absolving her from anything so obvious – or human – as self-interest.

Having received additional training in spinal disorders, the ESP was now sufficiently skilled to undertake 75 per cent of her consultant's traditional workload. Instead of routinely seeing all patients referred to him, and passing on a proportion to her to treat, he now merely scanned the referral letters, enabling him to concentrate on his surgical cases. This had already paid handsome dividends, with waiting times for a consultant appointment reduced to the national target of 14 weeks. Such was the apparent success of this new service model that it was to be rolled out nationally.

Observation in the ESP's clinic, however, revealed how incomplete even her detailed mapmaking and mapping were. For example, her map failed to capture the subtle differentiation that was being created in the process. The ESP's newly developed clinical skills elevated her above those nurse practitioners who, working tightly to protocol, assessed patients for arthritis of the knee and hip; 'That's a technicians' job' she explained. While enhancing her status and skills, and enriching her job, her new managerial and clinical responsibilities simultaneously served to differentiate her from her erstwhile colleagues:

> It's taken me away from working with my staff at the coalface, because I would have been in the Department treating patients alongside them. I think they regard me differently. I think they are almost in awe of me . . . Dealing with the turmoil and reaching a new position is what this is all about. The hierarchy means that if you're at the top, you've got the isolation of being at the top.

At the same time, while she cherished the opportunities for working more closely with her orthopaedic consultant, she did not mix with him socially. Intriguingly, she took the patient's seat at the afternoon's case discussion with him. Thus, as a member of 'a new species' the ESP found herself in a quandary, losing some of the camaraderie which she had previously enjoyed working as an 'ordinary' physiotherapist: she was betwixt and between (Pinder 2000).

Moreover, in adding texture to the map, she talked about the lack of a mentor, her own stress, and the difficulties of her secretarial staff in keeping up with her increased workload, highlighting the gap between her professional ambitions and the search for resources to nourish them. The calculus of costs and benefits was far from straightforward.

The pathways conveyed little sense of the disputed nature of the boundaries between professions. Each clearly marked box and confident arrow gave the impression that professional tasks and responsibilities were clearly delineated and could be precisely apportioned. A 'harmony ideology' (Nader 1990) was being presented, where each professional was working in concert with the others for the common good:

> There are no turf wars. We work as part of a team (Consultant dermatologist, Brown).

Pathways were perceived as functioning as tools for consensual decision-making:

> They give a structured frame-work; they're step-by-step plans to cover the things you should be generally thinking about (Nurse educator, Jameson).

Superficially, this dominant ideology of inter-professional harmony appeared to be confirmed by observation, which revealed several instances of team-work in practice, with the specialist ophthalmic nurse dovetailing neatly with ophthalmologists in the adjoining room, for example. However, other evidence suggested that the division of professional labour was not always as consensual or functional as the maps suggested, as the specialist dermatology nurse at Brown explained:

> Once you become reasonably confident in your role, when doctors have a busy clinic, they will off-load a little bit. The protocol says that all patients should be seen by a consultant, *then* by me. But I get asked, 'Can you see this patient?' What they forget is that I've never been a doctor, never been to medical school. A keratosis could be something more serious. But we're trying to find a way round that one. How to negotiate tactfully on that, otherwise mistakes will be made.

Moreover, while the additional professional space opened up by care pathways might benefit some practitioners within the profession concerned, it might be to the detriment of others:

> There might be a *little* bit of friction between those nurses who have not taken on that specialist role and those who have (Consultant dermatologist, Brown).

The optometrist, with an additional layer to his map (Figure 1, map (ii)) talked about the status differentiation between 'ordinary' optometrists and

> élite optometrists who won't work in the multiples, Boots, SpecSavers, the supermarkets. They really, really don't. Working for a multiple is considered to be the last thing you would ever do. It's both your hands and feet cut off working for a multiple.

Closer examination of the apparent orderliness and functionality of the pathway revealed a number of competing interests:

> they should all be working for the good of the patient instead of building their own little empires (GP specialist).

In similar vein, the specialist dermatology nurse in Brown spoke of 'Lines constantly criss-crossing all over the place' and of 'being taken over' by the newly appointed GP Specialist in Dermatology. Neither was the anticipated colleague-ship between hybrid professionals themselves much in evidence: the specialist ophthalmic nurse at Brown, for example, did not know the ESP only four doors away down the corridor.

Participants were aware of the ambivalent relationship between pathways and reality:

> I don't think they create reality because it's already there, all these things are already happening. What health service managers are trying to do is capture it, but they probably refine it too (GP specialist).

At the same time, he acknowledged their power to shape the reality they were supposed to be describing:

> It's a descriptive tool, describing variability and *attracting resources to it* (our emphasis).

Thus, the newly appointed dermatology consultant at Brown (discovered at interview, reducing a busy map to sleeker contours on the advice of the Chief Executive) was anxious to make a mark and reverse the dispiriting trail of ward closures for his specialty. He found the dermatology pathway to be an effective ally in arguing for additional resources:

> One of the difficulties is that I don't have my own outpatient clinic. I'm part of the general outpatient clinic, so that I share work and rooms with all the other specialists. I do one day in Clinic One, then another day in Clinic Two, and another day I'm somewhere else. That has the result of £11,000 of equipment being underused.

A follow-up interview revealed that this strategy had worked. An additional consulting room had been secured for his specialist nurse and a GP specialist in dermatology had also been appointed.

Pathways were not only a means of pinpointing where patients were in the system. They were also a way of keeping track of practitioners themselves. Formally, they allowed little latitude for either creativity or unruliness. Thus, the specialist dermatology nurse had less scope to exercise her clinical judgement, although she still prided herself on her 'holistic' approach, chasing the Accommodation Officer for parking concessions for her patients, for example: 'That's my style'. Similarly, while the ESP relished her new responsibilities, she remained subject to the disciplining gaze of her consultant, watchful for mistakes, and a rheumatology GP specialist sceptical of her ability to accurately decipher blood tests. Not surprisingly, some professionals were better placed than others to exploit what opportunities there were. A specialist GP explained her own back pain pathway, for example:

> When you're actually working, everybody has a pathway. Most doctors have a plan of what to do when they see someone with 'x', but it isn't necessarily set in stone . . . It's idiosyncratic. It should be standard but we all do different things. We've all got different interests. Ironing out

variations might actually bring the standard of care down. When I see someone with a bad back, I probably do more than some of my colleagues would because I'm interested in it – I've got other colleagues who are more interested in, say, asthma, so they would perhaps take more time with asthma patients. This is the thing, that's not a standard pathway that someone has told us to do. That's what *I* do.

Her independent contractor status afforded this GP a degree of autonomy of practice that was denied to others. Despite the overall shift towards standardised practice, freedom to be creative continued to be distributed along traditional hierarchical lines.

We have seen that the first section of the chapter points to the objectification of patients on the maps, glossing the tension between professionals' desire to respond individually to patients, and the difficulties of doing so. This section has illustrated how professional relationships were also objectified: the relationships that were idealised on the pathways as flow (the arrows) but not overlap (the boxes), actualised as well as reflected notions of appropriate professional distance. In theory, the maps represented them as consensual, orderly and discretely bounded. In practice, map-making and mapping revealed them to be both contested and imprecise.

Discussion

This chapter has drawn on the work of contemporary cultural cartographers to pose some critical questions about the burgeoning use of pathways in contemporary health-care policy. At the same time, we must make it clear that we do not oppose the agenda that informed them in our study sites. We were impressed by the efforts of our participants to reflect critically on their current practice, and to think systematically and holistically about services. We commend what appeared to be a genuine commitment to identifying 'gaps' in provision, removing 'bottlenecks' and smoothing the passage of patients through the system. As Corner (1999: 213) points out:

> We have been adequately cautioned about mapping as a means of projecting power-knowledge, but what about mapping as a productive and liberating instrument, a world-enriching agent . . . ?

Our concern is the modest one of drawing attention to the tensions endemic in pathways as in all 'rationalising techniques' (Berg 1997: 157). We simply urge awareness of the way they inevitably reflect and reproduce the very tensions (between efficiency and effectiveness, patient-centredness and organisational or political imperatives) that they strive to resolve. More specifically, are care pathways to be regarded as Taylorist devices for standardising care and treating each individual patient in precisely the same way? Or are they the means of affording individualistic treatment, while simultaneously creating organisational efficiency by 'tayloring' the organisation to the patient (rather than the other way round)[2]? Certainly, our respondents focused on the former interpretation rather than the latter. We, however, would argue that these tensions are inherent in any attempt to fix processes and meanings that are always open-ended and contingent. Hence our emphasis on the notion of a *process cartography* (Rundstrom 1993: 21), where the pathway is best seen as 'open-ended, on-going, always leading to the next instance of mapping'. As Ingold (2000) puts it, 'We know *as* we go, not *before* we go'.

We have seen that the care pathways we have been considering systematically omit the plasticity of patients' personal circumstances and lived experience, providing no map of the terrain that the ill person has to traverse. Thus, there is no space in them for the 'biographical disruption' (Bury 1982) entailed by chronic illness – the profound disturbance of taken-for-granted ways of thinking and behaving and relating to the world. Nor is there room for the subsequent task of biographical rebuilding (Siegel and Krauss 1991). Pathways overlook the problem, constant for people with stigmatised conditions such as epilepsy or HIV/AIDS, of 'managing the secret' (Caricaburu and Pierret 1995). Nor do they acknowledge that these considerations might influence patients' decisions about health service utilisation, or that services might possess important symbolic significance. Thus, people with HIV/AIDS might delay (or avoid) transfer to GP care because it signifies the transition from 'living with HIV' to 'dying from AIDS' (Petchey, Farnsworth and Williams 2000).

Hence, our study suggests that the empowering potential of pathways claimed by Parker (1999) is at best equivocal as far as patients are concerned. Although numerous policy documents emphasise the role of 'lay' people in decision-making, and the NHS Plan (DoH 2000) requires that 'patients must have more say in their own treatment and more influence over the way the NHS works', we found that a socially conservative vocabulary was being promoted. This paternalism was something which some professionals in the study were both aware of and uneasy about – and, to an extent, wanted to resist. However, whilst pathways enabled clinicians to spot-check where patients were in the system, it was questionable whether they shared this information with patients. Importantly, patients were not invited into the decision-making process: it was practitioners who had the power to make – or un-make – the map. Harvey's (1998) observation that 'Maps are pre-eminently a language of power, not of protest' applies equally to care pathways. The voices that they privileged were those of the experts – whether clinicians, managers or auditors (Hogg 1999, Berg *et al.* 2000). As usual, the voice of patients was muted.

As far as our participants are concerned, we must be careful not to paint too deterministic a picture, and this emerged most strongly in respect of the healthcare practitioners studied. Pathways were important mobilising metaphors, prescribing as well as describing, as the specialist ophthalmic and dermatology nurses and the ESP have illustrated. In practice, however, people don't do the ideal job: they do the do-able job. As embodied practices rather than simply 'things out there', they both reproduced the 'culture' of the NHS as well as offering scope for redefinition, but differentially along traditional hierarchical lines.

Like other protocols, pathways simultaneously shaped what was happening in practice, and the way practitioners thought about things, embodying and transforming clinical practice, at the same time as they were modified by it. As Berg has argued, contrasting the tool and practice (or map from map-making and map-ping) encourages advocates of the former to exaggerate its universality and to focus on 'resistance' to change; and its critics to view the tool as essentially misconceived and exaggerate the inherent superiority of local professional knowledge. Two dangers arise from this. The first is that of seeing *'blockages'* in terms of professional recalcitrance or personality flaws (the ENT consultant at Fleming), rather than the result of complex, interweaving socio-economic and political factors. In its turn, this may deflect attention away from more appropriate structural reform. Secondly, an obsession with dismantling 'restrictive practices' risks undermining the values which hold professional communities together. The subtle layering of hierarchy and responsibility as the new hybrid professionals become folded into an evolving system suggests that professional boundaries are not just bastions of rigidity, but important sources of identity and belonging.

As emphasised earlier, this chapter does not refute the good intentions and possible benefits of rational planning per se. Neither is it intended to be an exercise in cynicism. It does, however, question the apparent over-reliance by healthcare planners on pathways as a universal panacea (Bragato and Jacobs 2003). As our study has suggested, there are no cures to demand management that are free from side-effects; every attempt at clarity elicited further grey areas. As Scott (1998) notes:

> Any large social process or event will invariably be far more complex than the schemata we can devise prospectively or retrospectively to map it.

The findings point to the need to be flexible and approximate, rather than unbending and absolute, if only because the social circumstances which are routinely confronted are ephemeral, ambiguous and likely to constantly change. Policy-makers' desire to pin things down, on the other hand, may render practitioners incapable of responding imaginatively to the inevitable unpredictabilities of life.

Hence, our focus on cartography as process always in the making. A lighter touch than the confident images promoted by the pathway may be achievable, one less fervent in pursuit of those tangible outcome points which the pathway dangerously forecloses (Robinson 2001). As Huxley (1949) noted:

> The good life can only be lived in a society in which tidiness is preached and practised, but not too fanatically, and where efficiency is always haloed, as it were, by a tolerated margin of mess.

Notes

1 Known also as integrated care pathways, coordinated care pathways, critical care pathways, anticipated recovery pathways or care maps.
2 We are indebted to an anonymous peer reviewer for this point and this formulation.

References

Abbott, A.D. (1988) *The System of Professions: an Essay on the Division of Expert Labour.* Chicago: University of Chicago Press.

Agar, M. (1996) *The Professional Stranger: an Informal Introduction to Ethnography.* (2nd Edition) London: Academic Press.

Berg, M. (1997) *Rationalizing Medical Work: Decision-Support Techniques and Medical Practices.* Cambridge and London: MIT Press.

Berg, M., Horstman, K., Plass, S. and van Heusden, M. (2000) Guidelines, professionals and the production of objectivity: standardisation and the professionalism of insurance medicine, *Sociology of Health and Illness,* 22, 6, 765–91.

Black, J. (1997) *Maps and Politics.* London: Reaktion Books.

Borges, J.L. (1984) *Collected Fictions.* Translated by Andrew Hurley. London: Penguin Classics.

Bowker, G.C. and Leigh Starr, S. (2000) *Classification and its Consequences.* Cambridge: MIT Press.

Bragato, L. and Jacobs, K. (2003) Care pathways; the road to better care, *Journal of Health Organization and Management,* 17, 3, 164–80.

Bury, M. (1982) Chronic illness as biographical disruption, *Sociology of Health and Illness,* 4, 167–82.

Caricaburu, D. and Pierret, J. (1995) From biographical disruption to biographical reinforcement: the case of HIV positive men, *Sociology of Health and Illness,* 17, 65–88.

Campbell, H., Hotchkiss, R., Bradshaw, N. *et al.* (1998) Integrated care pathways, *British Medical Journal,* 316, 133–7.

Corner, J. (1999) The agency of mapping: speculation, critique and invention. In Cosgrove, D. (ed.) *Mappings.* London: Reaktion Books.

De Certeau, M. (1984) *The Practice of Everyday Life.* Berkeley: University of California Press.

Department of Health (1997) *The New NHS: Modern and Dependable. HMSO.*

Department of Health (2000) *The NHS Plan: a Plan for Investment, a Plan for Reform.* London: Stationery Office.

Etheridge, M.L. (1986) Timelines: the maps for managed care, *Definition,* Fall, 1–3.

Fayol, H. (1949) *General and Industrial Management.* London: Pitman.

Gellner, D.N. and Hirsch, E. (2001) *Inside Organizations: Anthropologists at Work.* London: Berg.

Hammer, M. and Champy, J. (1993) *Re-Engineering the Corporation: a Manifesto for Business Revolution.* London: Nicholas Brearley Publishing.

Harvey, J.B. (1992) Deconstructing the map. In Barnes, J. and Duncan, J. *Writing*

Worlds: Discourse, Text and Metaphor in the Representation of the Landscape. London: Routledge.

Harvey, J.B. (1998) Maps, knowledge and power. In Cosgrove, D. and Daniels, S. (eds) *The Iconography of Landscape.* Cambridge: Cambridge University Press.

Hogg, C. (1999) *Patients, Power and Politics: from Patients to Citizens.* London: Sage.

Huxley, A. (1949) *Prisons: the Carceri Etchings by Piranesi.* London: Trianon Press.

Ingold, T. (2000) *The Perception of the Environment: Essays in Livelihood, Dwelling and Skills.* London: Routledge.

King, G. (1996) *Mapping Reality: an Exploration of Cultural Cartographies.* London: Macmillan Press.

Lambert, H. and McKevitt, C. (2002) Anthropology in health services research: from qualitative methods to multi-disciplinarity, *British Medical Journal*, 325, 210–12.

Leicht, K.T. and Fennell, M.L. (2001) *Professional Work: a Sociological Approach.* London: Blackwells.

McNulty, T. and Ferlie, E. (2002) *Reengineering Healthcare: the Complexities of Organizational Transformation.* Oxford: Oxford University Press.

Nader, L. (1990) *Harmony Ideology: Justice and Control in a Zapotec Mountain Village.* Stamford: Stamford University Press.

NHS Information Authority, 2004. www.nelh.nhs.uk (visited 12 October 2004).

Norris, A.C. and Briggs, J.S. (1999) Care pathways and the information for health strategy, *Health Informatics Journal*, 5, 209–12.

Ouroussoff, A. (2001) What is an Ethnographic Study? In Gellner, D. and Hirsch, E. (eds) *Inside Organizations: Anthropologists at Work.* Oxford: Berg.

Owen, G. (1998) Extended scope practitioners in orthopaedic clinics: a growing field, *Rehabilitation International*, Fall, 33–8.

Parker, C. (1999) Patient pathways as a tool for empowering patients, *Nursing Case Management*, 4, 2, 77–9.

Pels, D. (2002) Everyday essentialism: social inertia and the 'Münchhausen effect', *Theory Culture and Society*, 19, (5–6), 69–89.

Petchey, R. Farnsworth, W. and Williams, J. (2000) The last resort would be to go to the GP. Understanding the perceptions and use of general practitioner services among people with HIV/AIDS, *Social Science and Medicine*, 50, 233–45.

Pinder, R. (2000) Betwixt and between: part-time GPs and the flexible working question. In Malin, N. (ed.) *Professionalism, Boundaries and the Workplace.* London: Routledge.

Power, M. (1997) *The Audit Society: Rituals of Verification.* Oxford: Oxford University Press.

Robinson, I. (2001) The virtues of never knowing: respecting process in health research. Paper presented at BSA Conference, York.

Rundstrom, R.A. (1993) The role of ethics, mapping and the meaning of place in relations between Indians and Whites in the US, *Cartographica*, 30, 21–8.

Scott, J.C. (1998) *Seeing Like a State: How Certain Schemes to Improve the Human Condition Have Failed.* New Haven and London: Yale University Press.

Siegel, K. and Krauss, B.J. (1991) Living with HIV infection: adaptive tasks of seropositive gay men, *Journal of Health and Social Behavor*, 32, 17–32.

Strathern, M. (ed.) (2000) *Audit Cultures: Anthropological Studies in Accountability, Ethics and the Academy.* London: Routledge.

Taylor, F.W. (1911) *Principles of Scientific Management.* New York: Harper.

Wolcott, H. (1999) *Ethnography: a Way of Seeing*. California: Alta Mira Press.

Woldrum, K. (1987) Critical paths: marking the course, *Definition*, 1–4.

Wood, D. (1992) *The Power of Maps*. New York: The Guilford Press.

Zander, K. (1991) What's new in managed care and case management? *The New Definition*, Spring/Summer, 1–3.

Chapter 6

Arguing about the evidence: readers, writers and inscription devices in coronary heart disease risk assessment

Catherine M. Will

Introduction

The *National Service Framework (NSF) on Coronary Heart Disease* (Department of Health 2000) set out comprehensive objectives for cardiology in the UK. The document's authors declared that 'identifying and treating those at greatest risk [was] one of the highest priorities of this NSF'. They asked general practitioners to develop systematic approaches to screening their patients, identifying modifiable risk factors and documenting advice or treatment for those at risk of heart disease, and recommended risk-assessment tools to help doctors perform these tasks. These tools compute data on physical characteristics to estimate the likelihood of an individual suffering a coronary event (or death) within a specified number of years.

Though equations for risk calculation in coronary heart disease have existed since the 1960s, they only attracted significant clinical interest in the 1990s. Since then the principles and practices of risk assessment have been hotly debated. Such debates echo those described by Marc Berg (1997) in his work on other decision-support technologies in medicine. Where Berg mainly uses ethnographies to reveal local negotiations surrounding the introduction of tools in clinical practice, I investigate a broader set of national and international arguments in the medical literature, focusing on reactions to three risk-assessment tools developed in the UK. I argue that published discussions and the design of the tools themselves are important interventions in wide-ranging debates about the character, cost and control of modern medicine.

Background to risk assessment in coronary heart disease (CHD) before 1995

The idea of characteristics that predict the 'risk' of someone suffering CHD has its own history, described by Robert Aronowitz in his book *Making Sense of Illness* (1998). Aronowitz argues that the Framingham heart study, begun in the US in 1948, was crucial for the development of the concept of the 'risk factor'[1]. Epidemiologists working on this study published the first CHD risk equation in 1967, using multivariate analysis of observed population data over eight years to create a function that could be used to predict disease in individuals (Truett *et al.* 1967). The equation was the focus of international disputes among epidemiologists, but received little attention from clinicians, especially in the UK.

In 1982 the World Health Organisation's guidelines on preventing coronary heart disease recommended that doctors 'stratify' their practice populations according to risk, to prioritise interventions. No suggestions were made about achieving this stratification. Guidelines from the Royal College of General Practice (RCGP) (1982) offered a simple checklist of risk factors. In 1986 epidemiologists at the London School of Hygiene and Tropical Medicine designed a risk scoring system to 'rank' patients more sensitively according to their CHD risk (Shaper *et al.* 1986). The score required the general practitioner to manually work through a formula with raw data on the patient, giving a score between 0 and over 1060. The Dundee risk disk (Tunstall Pedoe 1991) also scored patients and then converted this to a rank from 1 (high risk) to 100 (low risk).

The meaning of these risk-scoring systems changed amidst enthusiasm surrounding new drugs, HMG-Co-A reductase inhibitors or statins. These drugs were first licensed in the UK in 1989 to reduce cholesterol, understood as an important CHD risk factor. In the early 1990s large clinical trials began to indicate their value in secondary prevention of CHD (*i.e.* reducing future risk for people who had already suffered symptoms of the disease). Reports of the early trials focused attention on levels of cholesterol in the blood and appeared to have an important influence on clinical thinking. In 1992 the new RCGP guidelines (RCGP 1992) suggested that 'busy GPs' could prescribe statins according to 'levels of risk' defined by total blood cholesterol, 'high' risk being more than 7.8mmol/l, 'medium' risk at 7.0mmol/l or more and 'low' risk 6.5mmol or more[2].

In 1991, however, a new version of the Framingham risk equation had been published (Anderson *et al.* 1991). This differed from earlier versions in calculating risk for five or ten years rather than eight and including High Density Lipoprotein cholesterol values rather than total cholesterol alone. Although not without its critics, the equation was picked up by researchers around the world and built into a new generation of 'risk assessment tools'. These were designed to estimate risk with higher precision for an individual, working with a percentage probability over a given number of years, rather than a rank or score.

Theoretical framework

The convergence of the idea of risk factors, drugs to treat risk and more precise risk assessment tools clearly speaks to sociological debates about Beck's (1992) description of the risk society. Although heart disease might appear as a 'personal risk', CHD is often presented in professional and public settings as an epidemic in industrialised societies. Thus CHD could be understood as one of the 'global dangers' that shape the risk society.

Beck argues that such risks exist through the application of knowledge that attempts (but fails) to make them calculable. Risks 'require the sensory organs of science – theories, experiments, measuring instruments – in order to become visible or interpretable as hazards at all' (1992: 27).

In this chapter, I treat risk assessment tools as an emerging 'sensory organ of science'. I examine published material around three tools to build a case study of how risk may be perceived by professionals and how this perception can be made concrete in new technologies. The work of Marc Berg and Bruno Latour was particularly useful for this detailed reading of the tools and their presentation. Where Beck refers to 'sensory organs', Latour offers a more detailed description of 'inscription devices' in scientific practice, including experimental apparatus, imaging techniques and written conventions used in the creation of scientific 'data' from observation.

In *Laboratory Life*, Latour and Woolgar describe scientists as a 'tribe of readers and writers making use of a set of inscription devices' (1979: 69). The data produced by such devices are gradually collected and ordered into longer published accounts. Once published, they may be 'rejected, borrowed, quoted, ignored, confirmed or dissolved' by readers in the scientific community. Yet, from the 'smog' of half born statements, some are borrowed, reused and become uncontested, at which point 'a [scientific] fact has been constituted' (Latour and Woolgar 1979: 87).

The inscription devices in Latour's laboratory are themselves achieved 'stable' forms emerging from past controversies, although this is not central for his account. Berg's (1997) work focuses on periods of instability for devices in medical practice, reflecting a broader tendency for sociological research on medical technologies to look at their use within the 'embedded routines' and 'cultural contexts' of the clinic (Heath *et al.* 2003). Berg uses ethnographies of the development and introduction of systems to guide treatment of chest pain, abdominal pain and breast cancer in specific locations to argue that both tools and medical work are transformed in the process of introducing decision aids into clinical practice. He is influenced by actor network theory, which looks to the unstable origins of 'facts' and 'artefacts' in what has been called technoscience (Law 1991, Law and Hassard 1999). Other work in the sociology of health draws on this theory to consider the development of metered dose inhalers in asthma (Prout 1996) and advice on contraception (Dugdale 1999).

Such studies are important in demonstrating that the form and meaning of technologies may be contingent upon and contested among healthcare practitioners as well as scientists and designers. Unfortunately, there is limited evidence about the extent to which different risk assessment tools are in use in British healthcare settings. There is scope for work like Prior's (2001), which studied risk assessment technologies as inscription devices in a genetic clinic, revealing the complex 'web' of social, political, financial and scientific calculations around their local application. Yet we do have published material from debates about CHD risk assessment among medical and allied professions at a national (and international) level over about ten years, as different tools have been proposed, revised and reinvented in the pages of medical journals. The letters, articles and editorials informing this chapter can be seen as continuations of the 'conversations between scientists' that Latour overhears in the laboratory, but conversations in which a wider group is able to take part or listen. Doctors may be both readers and writers in these discussions, which reveal wide-ranging professional concerns about risk calculation in the clinic.

Berg (1997) argued that the context for the introduction of his decision to support technologies was a debate about whether medicine was 'art' or 'science', illustrating this argument with material from American medical journals after the Second World War. This chapter builds on the idea that the tools themselves are part of debates about the character of medicine, using a close examination of a specific British example. Berg's study covers several contrasting tools drawn from what he characterises as three principal types: statistical devices (including diagnostic tools and decision analysis); protocols; and expert systems. My focus is the ongoing competition between a set of tools with apparently similar aims. The 'risk assessment tool' might be described as existing in an ongoing state of indeterminacy, illustrated by the inclusion of five alternatives in the otherwise standardising NSF.

The Sheffield tables: an early attempt to guide statin use and manage the cost

The first risk assessment tools to gain prominence in the UK were the so-called Sheffield risk assessment and treatment tables (Figure 1). A team of clinical pharmacologists developed two tables, one for men and one for women. The earliest version was published in the Lancet in December 1995 (Haq et al. 1995) and was set to calculate 45 per cent risk of CHD death for an individual over ten years. The pictured table is the second, more commonly-cited version (in this case for men), which was published eight months later (Ramsay et al. 1996a) and was set to calculate 30 per cent risk of a CHD event. Other visual risk assessment tools had recently appeared in guidelines from New Zealand clinicians (Mann et al. 1993) and the European professional societies (Pyörälä et al. 1994). However, the Sheffield tables dominated discussions in the UK for several years. In 1996 they were distributed in a supplement to the Drug and Therapeutics Bulletin and formed the subject of a major debate at the British Hypertension Society conference that year. In 1997, the Standing Medical Advisory Committee to the government recommended the tables to all doctors in the National Health Service and copies were sent out to general practices (NHS Executive 1997). New versions of the tables appeared in 1998 and 2000, leading up to their inclusion in the NSF.

To use the table, doctors were asked first to read down to select a specific set of risk factors for an individual, including gender, smoking and hypertension. This set was read against approximate age on another axis, to

Men: Cholesterol concentration (mmol/L)

Hypertension	Yes	Yes	Yes	Yes	Yes	No	Yes	Yes	No	No	Yes	No
Smoking	Yes	Yes	No	No	Yes	Yes	Yes	No	Yes	No	No	No
Diabetes	Yes	No	Yes	No	Yes	Yes	No	Yes	No	Yes	No	No
LVH on ECG**	Yes	Yes	Yes	Yes	No	No	No	No	No	No	No	No
Age (years)												
70	5.5	5.5	5.5	5.5	5.5	5.5	5.5	5.5	5.5	6.0	6.5	7.7
68	5.5	5.5	5.5	5.5	5.5	5.5	5.5	5.5	5.6	6.4	6.9	8.2
66	5.5	5.5	5.5	5.5	5.5	5.5	5.5	5.7	5.9	6.8	7.3	8.7
64	5.5	5.5	5.5	5.5	5.5	5.5	5.5	6.1	6.3	7.3	7.8	9.3
62	5.5	5.5	5.5	5.5	5.5	5.5	5.6	6.5	6.7	7.8	8.3	
60	5.5	5.5	5.5	5.5	5.5	5.6	6.0	6.9	7.2	8.3	8.9	
58	5.5	5.5	5.5	5.5	5.5	6.1	6.5	7.4	7.7	8.9		
56	5.5	5.5	5.5	5.5	5.5	6.5	7.0	8.0	8.3			
54	5.5	5.5	5.5	5.5	5.9	7.0	7.5	8.6	9.0			
52	5.5	5.5	5.5	5.5	6.3	7.6	8.1	9.3				
50	5.5	5.5	5.5	5.7	6.9	8.2	8.8					
48	5.5	5.5	5.5	6.2	7.5	8.9						
46	5.5	5.5	5.5	6.8	8.2							
44	5.5	5.5	5.8	7.4	9.0							
42	5.5	5.6	6.4	8.2								
40	5.5	6.1	7.1	9.0								
38	5.5	6.8	7.9									
36	6.0	7.6	8.8									
34	6.7	8.6										
32	7.6											
30	8.7											
<29												

Figure 1 *Sheffield table for primary prevention of coronary heart disease. Showing serum cholesterol concentration conferring an estimated risk of coronary events of 3.0 per cent per year (or 30 per cent over ten years).*
Source: Ramsay *et al.* 1996a

take the reader to the main body of the tables. Here a value for total cholesterol was included in approximately half the cells. If a cholesterol level was included, the user was advised to perform a cholesterol test on the patient's blood. The result of this test should again have been read with reference to the table. If the blood proved to contain cholesterol equal to or greater than that indicated in the table, the individual was said to have a 30 per cent or greater risk of suffering a coronary event over ten years. Where cholesterol was omitted after the initial risk factors had been read off, the absence indicated that no value of total blood cholesterol could produce a risk as high as 30 per cent (with the existing combination of other factors).

The text accompanying the published tables advised that where risk was 30 per cent or more, statin treatment should be initiated. This marked a change from earlier scoring systems, which had made no suggestions about treatment. The tables were promoted as a tool to target expensive HMG-CoA reductase inhibitors – rationing was clearly part of their role, although the actual word was not used. The second Sheffield tables were produced with a reduced risk threshold after the publication of the West of Scotland Coronary Prevention study (Shepherd *et al.* 1995), which indicated that statins lowered cholesterol and the risk of cardiovascular disease before symptoms were apparent (primary prevention). However, there was anxiety within the Department of Health that the cost of statins could cripple the NHS if doctors treated everyone who might benefit according to the trials.

The Sheffield tables should thus be seen as interventions in a debate about statin costs: in the authors' words as 'the key to rational use of HMG-CoA reductase inhibitors' (Haq *et al.* 1995). As we have seen, previous guidance on using statins in the UK had focused on cholesterol levels. By asking doctors to treat 'global risk', the table acted as a gatekeeper not only to expensive statins but also to the cholesterol test itself. However, even as the tool made this attempt, cholesterol was visually placed in the foreground and was the focus at the moment of the treatment decision.

The Sheffield table was not simply a tool to restrict GPs' prescription of statins in general. A specific assumption about the ability of the NHS to pay was built into the design as the authors claimed that the risk threshold was guided by estimations both of numbers likely to qualify for treatment ('people' who would become 'patients') and how many statins the UK could afford. This calculation was openly debated and discussed in response to the Lancet articles. However, in the table itself, as published and sent out to GPs, the figure of 30 per cent was not directly stated. Instead, it remained implicit, and later became a 'standard' of a kind, reappearing in the NSF recommendation to treat 30 per cent risk five years later, detached from any discussion of cost and any particular tool.

Looking at the table as a visual object, it is apparently simple and relatively plain. Its form seems to make a modest claim: to carry epidemiological data into the consultation to support clinical judgement about a person's risk status and treatment, a moment of diagnosis and prognosis combined. The table, however, also worked to guide a set of medical practices for doctors: requiring data from tests and questions, reiterating the risk factors deemed relevant, guiding decisions about further tests and initiating drug therapy. The table was thus a kind of concertina of protocol, encoding what Latour has called a 'programme of action'. This stretching of the meaning of the tools may upset Berg's analytic division between computer inference tools and protocols.

The Sheffield tools were closely identified from the start with the emerging evidence-based medicine (EBM) movement. In the 1995 *Lancet*, Rodney Jackson (of the EBM Unit in Oxford) and Robert Beaglehole (1995), announced the arrival of the tables in glowing terms, declaring that such quantitative assessment of pre-treatment prognosis was an important but all too often neglected element of EBM. This comment, the article introducing the table and early responses were indexed in the *Lancet* in the category 'evidence-based medicine' (only previously used on two occasions, in 1994) rather than under 'general practice' or even 'risk,' though these categories were also available. This connection supports Berg's suggestion that the tools may be used to express and refine a particular view of medicine.

Critical conversations begin: printed responses to the Sheffield tables

The debates following the Sheffield publication rehearsed key issues in the evidence-based approach and the tools' design. These can be summarised as having three themes – the nature and limitations of the evidence used to build the tool, the place of cost calculations and the 'useability' of the tools in clinical practice. Here I draw on a representative selection from a huge number of letters, editorials and commentaries appearing in *The British Medical Journal* and *The Lancet*[3].

At first, the Sheffield tables were attacked on the selection of risk factors and how they were featured. In a letter to *The Lancet*, Drs Wierzbicki and Reynolds, chemical pathologists, complained that the earliest tables did not include any reference to hypertriglyceride in the blood or high density lipoprotein (HDL cholesterol) concentration, which Framingham had shown to be good predictors of CHD at the population level (Wierzbicki and Reynolds 1996a). The authors accepted the comment, and the second version was developed using a population mean HDL for men and women at different ages. The same critics complained that the dichotomous 'hypertension' variable was inadequate – the incidence of heart disease had been shown to increase in line with continuously increasing blood pressure levels. Saying 'yes' or 'no' to a single level (in this case fixed by the British Hypertension Society) would reduce the accuracy of the prediction. However, the authors defended this step. 'Dichotomising systolic blood pressure does sacrifice accuracy for the sake of simplicity', they wrote, but made the tool easier to use. Although they were working on including the ratio of total cholesterol to HDL, 'we are uncertain whether the resulting table would be as acceptable to ordinary doctors – an overriding concern for us' (Ramsay *et al.* 1996b).

The authors of the Sheffield tables were willing to compromise on accuracy to increase the accessibility of the tool. This could be difficult, given their desire to base the tool firmly on 'evidence'. Problems arose where there were limited data from trials or studies. Jackson and Beaglehole (1995) themselves acknowledged the difficulty of restricting guidance to patients under the age of 70, as no randomised controlled trials had yet included older patients. The greatest debate occurred around the precise degree of risk being targeted in the tables, which one GP (Jones 1997) complained had been chosen 'arbitrarily'. Wierzbicki and Reynolds (1996b) asked why the tables did not act on clinical evidence for benefit at 1.86 per cent risk per year in participants in the West of Scotland study. The problem exposed tensions in EBM which sought both to act on 'scientific' evidence, typically the randomised control trial, and to allocate treatment rationally with due consideration of comparative costs. Rodney Jackson agreed that more debate was needed, but argued that this might be done by comparison of cost per life year gained from this and other interventions in CHD and beyond. Yet the authors defended the decision to use the national cost calculation in its own terms (Ramsay *et al.* 1996b). Using the West of Scotland level would entail treating about 25 per cent of the adult English population, they claimed, and would cost the UK government more than £3 billion per annum.

A similar argument was used to answer a Taunton GP who was concerned that treating those with greatest absolute risk meant that younger patients were denied treatment when they had more potential years of life to gain (James 1996)[4]. Haq *et al.* (1996) agreed that it was desirable to treat younger groups, but insisted that this was 'beyond reach at present' based on current UK prices. However, Wierzbicki and Reynolds (1997) maintained their critical stance, arguing in the *BMJ* that cost should not be given so much weight and pharmaceutical prices could perhaps be renegotiated. Given these complaints about the way in which the Sheffield tool might ration treatment, it is ironic that the tool's authors attracted funding from pharmaceutical companies to promote the tables in these early years, as the companies thought they would actually increase statin prescription in the UK by improving the drug's visibility.

The Standing Medical Advisory Committee quickly defended its endorsement of the Sheffield tool in the *BMJ* (Enoch 1997) following the Sheffield team's appeal to simplicity. Cost might change and there were questions about the tables' accuracy but 'most statins will be prescribed by general practitioners, and it was for them that authoritative but concise interim advice was primarily needed' (1997: 1615). In response the critics gathered the signatures of '103 professors, consultants and specialists in preventive cardiology, chemical pathology, metabolism and lipids, clinical pharmacology, epidemiology and public health' on a letter expressing the concern that the advice was too concise and that 'doctors [probably meaning their colleagues in general practice] will use the table to avoid cholesterol testing in younger patients without noticing the "small print"'' (Reynolds *et al.* 1997). Despite suggesting in an interview that they had 'killed off' the Sheffield tables with this letter, Reynolds and Wierzbicki continue to include the tables in their comparative studies of risk assessment methods.

'A medical response' in the form of a new tool: Joint British Guidelines and the quest for a professional consensus

By 1997 there was some agreement that risk assessment tools had to negotiate a balance between simplicity and accuracy, a formulation leaving out the question of cost. Disagreements remained and a new tool emerged as a challenge to the Sheffield tables in guidelines from a joint committee of the British Hyperlipidaernia Association, the British Hypertension Society and the British Cardiac Society (Wood *et al.* 1998). The tool was heavily based on the New Zealand charts, and for a time the discussion was couched as a competition between these and the Sheffield tables. In a series of letters Poulter *et al.* (1998) argued that the New Zealand tools gave more scope for the clinician: they did not 'pre-empt' cholesterol measurement (Poulter *et al.* 1997) and had less 'didacticism'. Although one of the Sheffield team, Larry Ramsay, was an adviser to the Joint Societies, in later comparative research the different tools were often portrayed as rivals (*e.g.* Ritchie and Isles 1999, Durrington 2000), reflecting professional differences within the guideline committee. This continued discussion might be seen as a sign of failure for the Joint British Societies, which had hoped to end debate with their consensus statement.

The new tables (Figure 2) were designed in Manchester and were endorsed after publication by the British Heart Foundation, who paid for copies to be provided to every GP in the UK. The guidelines were several pages long and included details on acquiring a piece of 'risk calculation software' as well as visual tables on paper. The tables were provided for those GPs who did not yet have access to a computer. In theory, by 1998, this would be a relatively small group, yet in practice the tables were more often referenced and discussed than the software alone[5]. On paper, the tables (Figure 3) divided the population into groups by age, gender and other characteristics, each with their own charts. This allowed them to have blood pressure as a continuous variable, where Sheffield had used a dichotomous variable for hypertension. It also meant that the user hunted for a chart corresponding to a particular *individual* (a male, smoker, diabetic etc.). Cholesterol was included on the second axis, expressed as a ratio of total blood cholesterol to High Density Lipoprotein. This served to direct attention away from any single cholesterol value in itself.

The Joint British tables also attempted to distance themselves from the political calculation embedded in the Sheffield tool. Risks of suffering CHD events of 30, 20 and 15 per cent were shown with curved sections on the graph, although there was now trial evidence for benefits from statins in those with risks as low as six per cent over ten years. Where the European guidelines had used 20 per cent risk, British doctors were encouraged to start only with those at 30 per cent risk but in practice to treat as many people as possible. However, in the text it was noted that until all those at 30 per cent risk had been picked up, it did not seem worth pursuing other groups. Thirty per cent was now seen as an achievable target for clinical intervention. Scientific evidence, however, was rather carefully distinguished from evidence on cost effectiveness: 'the approach employed is not always strictly evidence based . . . both the volume of evidence available and the pressure on resources that would follow its widespread uncritical adoption necessitates caution' (Wood *et al.* 1998).

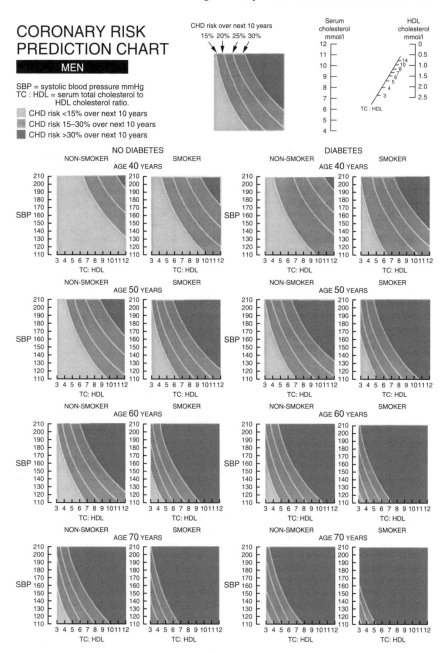

Figure 2 *Joint British Societies coronary risk prediction chart for men*
Source: Wood *et al. Heart* 1998

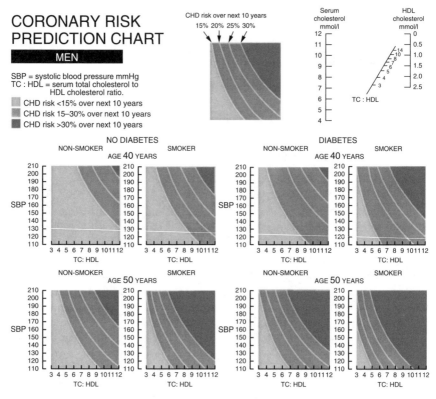

Figure 3 *Joint British Societies coronary risk prediction chart for men (detail)*
Source: Wood *et al. Heart* 1998

In several ways the Joint British tables might be seen as a defence of professional authority against government restrictions or pharmaceutical marketing (*c.f.* Armstrong 2002). The authors argued that the tables were an appropriate *medical* response to the problem of whom to treat with statins. 'As physicians we are dealing with the whole person' (Wood *et al.* 1998: S1) they declared, and suggested that calculating total risk to the individual was a great improvement on a simple cholesterol test, which might be done by a health insurance or occupational health organisation. While they were filling a gap left in past risk perceptions, they did not offer strict guidelines on when to prescribe, leaving space for something understood as clinical judgement. Nonetheless, like the Sheffield table, these tools demanded certain actions from the doctors using them: both needed a full complement of data to function[6]. These demands were more explicit than before, detailed for readers of the accompanying text with the full weight of combined professional conviction.

Many ordinary doctors were still unhappy or confused. In 2000, the *BMJ* published an entire issue on risk calculation, which included the latest versions of both the Sheffield tables and a summary of the JBS guidelines.

Several GPs expressed continued concern about the question of age. Dr Daniel Albert from Leeds Primary Care Trust combined the problems of applying the charts and their management of statin costs in a strongly worded critique of the whole enterprise:

> To use risk prediction charts or computer programs to bring everyone's risk down to the same level will maximise the number of lives saved but not the amount of life saved. It will provide quick results to please NHS planners and will divert much NHS funding into selected pharmaceutical companies, but the impact will be on those near the end of their lives anyway. It will have relatively little effect on the epidemic of people dying in middle age from heart disease (Albert: 2000).

Other comments returned to the balance between simplicity and accuracy. One letter referred to research that had suggested that the latest Sheffield tables were difficult to use (Durrington 2000). Using the entire Framingham risk equation would appear to provide the most 'accuracy' – but this also seemed cumbersome in practice. The JBS tried to solve this by providing the equation on disc and some doctors apparently used this in spreadsheets or even hand held computers to perform the function, which was also included in the new EMIS software for general practice[7]. Others worked to 'package' the equation to improve 'useability' for the general practitioner (Vallance and Martin 1997). These efforts resulted in the final tool to be discussed in this chapter.

Another clinical pharmacological intervention: the UCL tool and shaping the medical consultation

Like the Sheffield tables, the final tool (Figure 4) was produced by clinical pharmacologists, in a team at University College London (Hingorani and Vallance 1999a). It was published in the *BMJ* and the software was available from *BMJ* Publishing, priced £60 (all the other tools had been provided free on request). The tool was designed to be loaded onto a GP's computer, and presented them with a template into which data on risk factors must be entered. The software then performed calculations to predict absolute risk of both coronary events and stroke.

The programme was promoted as being 'both simple and flexible'. The authors admitted that it was limited by the data used to build it (the Framingham function), like other tools. It might not deal well with patients with diabetes or familial hypercholesterolaernia who had formed very small proportions of the Framingham populations. Arguing, however, that 'practising clinicians are [also] presented with the published data, not raw individual data', the UCL team insisted that 'the program is similar to current practice but has the advantage that it applies the results in a consistent manner with a perfect memory' (Hingorani and Vallance 1999a: 104–5). Once again the tool was presented as fitting current practice, not transforming, replacing or improving on judgement.

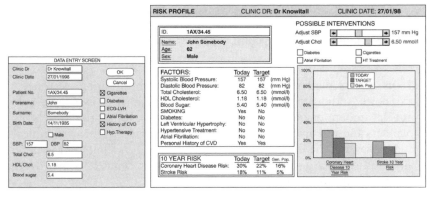

Figure 4 *Data entry screen and individual risk profile for computer program for guiding management of cardiovascular risk factors and prescribing*
Source: Hingorani and Vallance *BMJ* 1999

The design of the UCL tool suggested that it would have an even greater impact on the shape of the consultation than other tools. Boxes allowed data on risk factors to be entered, stored and displayed along with the risk calculation and a graphical representation (Figure 5). Coloured graphs were apparently intended to make information on 'your risk' available to patients. These also allowed comparisons with someone of the same age and gender in the general population (relative risk). Continuing the focus on reducing risk, sets of targets were suggested, to 'improve' both individual risk factors and overall risk, although this relied on the assumption that this person, without risk factors such as smoking or high blood pressure, would in future have the same, lower risk as someone who had never smoked or suffered from hypertension. The accompanying paper referred to the 'problems facing doctors to implement the findings of the many clinical trials and cohort studies into everyday clinical practice *and* how to involve patients in the decision-making process' (Hingorani and Vallance 1999a: 101) and with these graphs and targets the tool made more convincing efforts than previous ones to help doctors achieve the second goal.

In addition to its other functions, 'the program should enable patients to make informed decisions about which interventions they would wish to pursue' (Hingorani and Vallance 1999a: 104). However, in the published article, the presentation of interventions appeared to be aimed more at the concerns of GPs (Figure 6). Graphs showing different 'treatment packages' claimed to aid a decision about overall response, not just whether to prescribe statins. These showed likely risk reductions from different interventions, but also indicated the direct costs of different interventions based on cost to the fund-holding practice, in contrast to the macro-economic cost calculation embedded in the Sheffield table. The paper spoke further to the imagined concerns of the entrepreneurial fund-holding GP suggesting that the tool

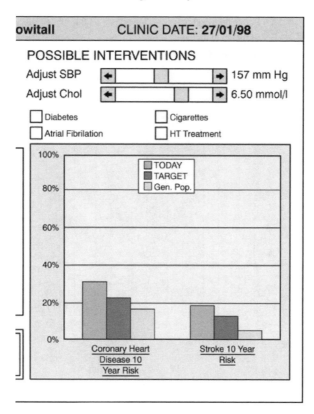

Figure 5 *Individual risk profile for computer program for guiding management of cardiovascular risk factors and prescribing (detail)*
Source: Hingorani and Vallance *BMJ* 1999a

could help the practice by storing data for audit or prescribing checks. This presaged the inclusion of calculators in the NSF, where they were arguably as much about encouraging data collection and facilitating audit on a national scale as directing clinical practice.

The UCL tool was apparently well received by general practitioners, judging by the responses and letters published in the *BMJ* – although these continued now familiar debates about the problem of access and benefit to treatment for different ages and patient groups. While the evidence base for the risk-factor intervention appeared less controversial, a new angle appeared due to the tool's effort to show risk reduction with different interventions. A major debate ensued about the validity of this. Several letters agreed that showing hypothetical risk reduction might have a place as a 'pedagogical' approach, 'as a way of involving individuals in the treatment decision', but warned that the effect of changing risk factors was unclear, and that such advice might be dangerous and misleading (Bartlett 1999, Gueyyffier 1999). The tool's authors however expressed themselves as satisfied:

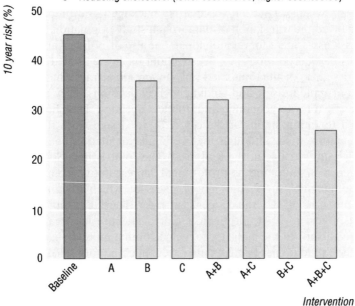

A = Reducing blood pressure (lower cost £1.31, higher cost £25.40)
B = Stopping smoking
C = Reducing cholesterol (lower cost £16.65, higher cost £33.30)

Figure 6 *10 year risk for individual and predicted risk reductions from interventions assuming reduce systolic blood pressure by 20mm Hg and total cholesterol by 20 per cent*
Source: Hingorani and Vallance *BMJ* 1999a

In designing this program, our aim was to take forward theoretical discussions about multiple risk factors, risk calculations and the implementation of thresholds for statin and anti-hypertensive treatment. In particular, we felt there was a need to develop a tool which would aid the practising doctor and 'operationalise' the extensive available evidence on risk factors and intervention. The level of interest which our paper has generated, particularly inquiries about the availability of the program, seems to endorse our view that there is a great clinical need for a program such as ours (Hingorani and Vallance 1999b).

Discussion

In the quote above, the designers of the UCL tool expressed an interest in stimulating debate as much as fulfilling a clinical need. Though some tools have become part of clinical work, discussion about the merits of several devices has continued in the pages of the medical journals where all the tools were first presented, as well as at conferences and other professional venues. No single tool can yet be said to dominate or to have achieved 'taken-for-granted' status in practice or publication. Yet several tools have achieved recognition on a national, even international, stage. To use Latour's language, epidemiological knowledge has been successfully packaged and prepared to travel. This involved the translation of data from a study population (Framingham) into a risk assessment for an individual, but also of political and professional concerns beyond positions in a debate about medicine as art or science.

The designers of the tools tried various ways to attract allies and therefore succeed in standardising and guiding medical decisions. They also responded in different ways to economic and political agendas. The Sheffield tool put the need to ration statins at the centre of its design, with the 30 per cent treatment threshold, although the economic calculation about cost to the nation was not visible in the tool itself. The Joint British Societies claimed less interest in rationing as professional representatives, but continued to make 'pragmatic' suggestions about levels at which doctors might offer treatment (now 30, 20 and 15 per cent risk). The UCL tool revealed little interest in cost in itself, although it hinted at more local calculations and suggested a shift in the audience to include both a more fleshed-out image of the general practitioner *and* patients. Over time, the tools also became more visually complicated, as if users were likely to accept and understand more detailed devices. Despite such differences, the designers agreed that they all had to balance 'acceptable accuracy' with simplicity and to use several common techniques for this accommodation.

Despite admitting that a compromise had been made, the accuracy of the tools was implied in their visual staging. They all relied on the representational devices of science such as graphs and tables. Features such as percentages, spatial categorisation and continuous variables helped convey prevision and certainty. However, the actual epidemiological and pharmacological data built into the tools was elided. The doctor never makes the actual mathematical calculation of the Framingham risk equation. The local and national cost estimates were rendered invisible, especially in the case of the Sheffield tables. The devices are by their nature highly indexical, but no less scientific for that. Latour (1987) has argued that 'science is characterised neither by an ability to escape indexicality nor by an absence of rhetorical or persuasive devices'. The technology of CHD risk assessment is similarly defined by a complex interplay between the indexical and rhetorical, between things carefully hidden and deliberately displayed.

Although the equation behind the risk calculation was black-boxed, the idea of steps towards assessment was in some sense opened out. The user is taken through several stages towards a judgement, in a spatial and temporal rehearsal of the key risk factors and their combination. In both the Sheffield and the Joint British Societies' tools the writers explicitly compromised to make this visual play possible, by making hypertension a dichotomous variable or by using wide age categories, while the UCL tool used a computer interface to manipulate a full set of data. The papers presenting all the tools persisted in explaining their design and were often open about choices (*e.g.* economic) that are left implicit in the actual tools. If we consider these papers and subsequent correspondence alongside the tools themselves, it seems that doctors were actually offered a series of educative moments about CHD risk, the techniques of using evidence in medicine and the types of decision-making this required.

It is possible to see the tool designers as inhabiting the more agonistic worlds of science than of pragmatic worlds of clinical practice. Two of the tools were designed in university departments of clinical pharmacology for use in specialist clinics. The Joint British Societies' tables also originated in the university and gained backing from élite clinicians who were heavily involved in research. In some ways their interest in 'simplicity' should alert us to the fact that they always promoted tools to *other* groups of doctors. All the designers were aware of a clinical audience, who had to be appealed to and educated. I would argue that they did not intend consumers to use a tool without understanding its limitations. In the medical journals, readers were given material to assess the evidence for themselves, and were asked to tolerate knowing that none of the tools could be perfectly 'accurate' in the assessment of CHD risk for an individual. As arguments about the use of evidence were rehearsed, they involved a widening group of doctors, creating agents capable of evaluating the tools and practising the kind of medicine that they presupposed.

The tools and this 'education' were not received passively. In this chapter, I have presented only a small sample of critical responses, which increasingly came from generalists as well as specialists. GPs may have complained of being confused by the variety of technologies on offer, but were also quick to learn how to attack the tools. The resulting debates became, in the words of one respondent, 'totemic' for the EBM movement. They are a site where this movement has been promoted, attacked and defended, but they increasingly drew even critics into using the language of EBM to make their case. The introduction by Jackson and Beaglehole (1995) was already significant in suggesting that 'the Sheffield risk and treatment table will be a useful tool for clinicians who are unfamiliar with the concept of absolute risk, but we believe they will soon outgrow it'. The tools were used in turn to help doctors master a set of concepts around CHD prevention, the relevant risk factors and comparisons between absolute and relative risk assessments. This project may have been more successful than intended. Recent interviews with tool designers have revealed a new theme: since GPs rarely apply the tools in an even-handed manner, nurses are likely to be more docile future consumers.

Conclusions

Risk assessment tools in CHD are significant as hybrid expressions of professional, political and economic agendas around the use of statins in UK clinical practice. Although there was some doubt, the most recent professional guidelines on hypertension and cardiovascular disease have retained risk assessment tools (new versions of the JBS tables) and used risk thresholds to structure treatment advice (Williams *et al.* 2004). The references to the tools by the SMAC and the NSF appeared to bolster their claims to offer a 'rational' approach to CHD prevention. It is not clear what weight to give such bureaucratic support or professional promotion, particularly in the case of the Joint British Societies, which may claim to be fighting bureaucratic interference. There is scope for further research on the impact of such endorsements on routine clinical practice, but even without this, the tools and discussions around them are important interventions in debates about statins, drug rationing and the use of evidence in medicine.

Latour's theories of hybrid networks constituted by alliances offer potentially strong models to investigate these problems further, just as his theory of 'transformation through translation' sensitises us to the varied and changing meanings of the risk assessment devices. My interest is in the multiple projects or networks into which the 'generic' risk assessment tool has been recruited: not only influencing treatment, but also the collection and interpretation of new data in CHD prevention and attempts to standardise care. Only the UCL tool makes any effort to involve the patient, but this is a theme that has attracted more attention recently. Doctors remained the imagined consumers throughout the debates examined in this chapter, so that the issue is not only what makes a *good enough* risk tool at a particular time in British cardiology, but also how negotiations of that point engaged and educated doctors about the limitations of these devices and the meaning of their own role.

Acknowledgements

I am particularly grateful to all the people who agreed to be interviewed on the subject of risk assessment in coronary heart disease. I also wish to thank Professor Joan Busfield, Dr Pamela Cox, audiences at the Universities of Essex, Newcastle and Nottingham and the BSA Medical Sociology Conference, as well as the two anonymous reviewers, for helpful comments on earlier versions of the paper. This research was funded by a doctoral grant from the Economic and Social Research Council.

Notes

1   The Framingham Heart Study was a major piece of observational research, which tracked the prevalence of various suspected 'risk factors' and of cardiovascular disease in several thousand American men and women. Beginning in 1948, it eventually included three generations and continues to be an important source for cardiovascular epidemiology today. Some more recent risk assessment tools have been built using data from clinical intervention trials, which show CHD morbidity and

mortality for patients who have received treatment for factors such as hypertension and cholesterol, and which may refer specifically to European or national populations. Framingham remains however a widely acknowledged 'gold standard'.

2 Mmol/L refers to millimol per litre.

3 Between 50 and 150 'citations' can be found on Web of Science of each of the three papers first introducing the tools in published form. However, such figures are unreliable as different journals had different and changing policies about how to count things like letters, rapid responses and even editorials. My examples are drawn from a qualitative analysis of two major journals.

4 From trial data, statins appeared to reduce risk by about a third in all patients, whatever the original cholesterol level. Those with highest absolute risk to start with therefore also saw the highest risk reduction – so their treatment was seen as most cost effective. Inevitably this also meant treating more older people, as age was a major component of the risk assessment.

5 Research in 1994 suggested that 79 per cent of practices were computerised, 63 per cent using them in consultation (Anonymous (BJGP) 1994). However, it seems that this IT use was often limited to prescription.

6 Only one tool does not require the full set of data – the inclusion of the Framingham equation on EMIS software allows for the use of 'average' mean values for the population for risk factors on which there was no information for the individual.

7 Some commentators have suggested that this should not be seen as improving the accuracy as the equation itself has error for each factor (thanks to both measurement error and biological variation), and the equation itself often has wide confidence intervals. Calculating the full equation from software also risks confusing treated and untreated values: the Framingham equation is based on untreated values only.

References

Albert, D. (2000) Absolute cardiovascular risk is not most appropriate measure to use (Letter), *British Medical Journal*, 321, 175.

Anderson, K.M., Odel, P.M., Wilson, P.W.F. and Kannel, W.B. (1991) Cardiovascular disease risk profiles, *American Heart Journal*, 121, 293–8.

Anonymous (1994) Editorial: communication in the year 2000, *British Journal of General Practice*, 44, 387, 434–5.

Armstrong, D. (2002) Clinical autonomy, individual and collective: the problem of changing doctors' behaviour, *Social Science and Medicine*, 55, 1771–7.

Aronowitz, R. (1998) *Making Sense of Illness: Science, Society and Disease*. Cambridge: Cambridge University Press.

Bartlett, W.A. (1999) Computer programs for guiding management of cardiovascular risk factors (Rapid response to Hingorani 1999), *British Medical Journal*, eletters consulted online at http://bmj.bmjjournals.com/cgi/eletters/318/7176/101

Beck, U. (1992) *Risk Society: Towards a New Modernity*. London, Newbury Park and New Delhi: Sage.

Berg, M. (1997) *Rationalizing Medical Work Decision-Support Techniques and Medical Practices*. Cambridge MA. and London: The MIT Press.

Department of Health (2000) *National Service Framework for Coronary Heart Disease. Modern Standards and Service Models*. London: HMSO.

Dugdale, A. (1999) Materiality: juggling sameness and difference. In Law, J. and Hassard, J. (eds) *Actor Network Theory and After*. Oxford: Blackwell Publishers.

Durrington, P.N. (2000) Joint British Societies recommend their computer program for risk calculations, *British Medical Journal*, 320, 660–1.

Enoch, P. (1997) Use of statins. Adequacy of SMAC's statement should be judged by clinicians, not health economists, *British Medical Journal*, 315, 1615.

Gueyyffier, F. (1999) Hypothesis of the program is flawed, *British Medical Journal*, 318, 1418.

Haq, I.U., Jackson, P.R., Yeo, W.W. and Ramsay, L.E. (1995) Sheffield risk and treatment table for cholesterol lowering for primary prevention of coronary heart disease, *Lancet*, 346, 1467–71.

Haq, I.U., Jackson, P.R., Yeo, W.W. and Ramsay, L.E. (1996) Sheffield risk and treatment table for primary prevention of coronary heart disease (Authors' reply), *Lancet*, 347, 468–9.

Heath, C., Luff, P. and Sanchez Svensson, M. (2003) Technology and medical practice, *Sociology of Health and Illness*, 25, 75–96.

Hingorani, A.D. and Vallance, P. (1999a) A simple computer program for guiding management of cardiovascular risk factors and prescribing, *British Medical Journal*, 318, 101–5.

Hingorani, A.D. and Vallance, P. (1999b) Authors' reply, *British Medical Journal*, 318, 1418.

Jackson, R. and Beaglehole, R. (1995) Commentary: evidence based management of dyslipidaemia, *Lancet*, 346, 1440–1.

James, M.A. (1996) Sheffield risk and treatment table for primary prevention of coronary heart disease (Letter), *Lancet*, 347, 466.

Jones, A.F. (1997) Statins and hypercholesterolaernia: UK Standing Medical Advisory Committee guidelines (Letter), *Lancet*, 350, 1174–5.

Latour, B. (1987) *Science in Action*, Cambridge MA.: Harvard University Press.

Latour, B. and Woolgar, S. (1979) *Laboratory Life: The Construction of Scientific Facts*. Beverly Hills and London: Sage Publications.

Law, J.A. (1991) *Sociology of Monsters: Essays on Power, Technology and Domination*. London and New York: Routledge.

Law, J. and Hassard, J. (eds) (1999) *Actor Network Theory and after*, Oxford: Blackwell.

Mann, J.I., Crooke, M. and Fear, H. *et al.* (1993) Guidelines for detection and management of dyslipidaemia, *New Zealand Medical Journal*, 106, 133–42.

NHS Executive (1997) *Standing Medical Advisory Committee Statement on the Use of Statins*, Executive Letter EL, 97, 41, Wetherby, West Yorkshire: Department of Health.

Poulter, N.R., Alberti, K.G.M.M., Beevers, D.G., Betteridge, D.J. and Campbell, R.W.F. (1997) Drug therapy for coronary heart disease: the Sheffield table (Letter), *Lancet*, 350, 1852–3.

Poulter, N.R., Alberti, K.G., Beevers, D.G., Betteridge, D.J. and Campbell, R.W. (1998) Drug therapy for coronary heart disease: the Sheffield Table (Letter), *Lancet*, 351, 443–4.

Prior, L. (2001) Rationing through risk assessment in clinical genetics: all categories have wheels, *Sociology of Health and Illness*, 25, 5, 570–93.

Prout, A. (1996) Actor-network theory, technology and medical sociology: an illustrative analysis of the metered dose inhaler, *Sociology of Health and Illness*, 18, 2, 198–219.

Pyörälä, K., De Backer, G., Graham, I., Poole-Wilson, P. and Wood, D. (1994) Prevention of coronary heart disease in clinical practice. Recommendations of the Task Force of the European Society of Cardiology, European Atherosclerosis Society and European Society of Hypertension, *European Heart Journal*, 15, 1300–31.

Ramsay, L.E., Haq, I.U., Jackson, P.R., Yeo, W.W., Pickin, D.M. and Payne, JN. (1996a) Targeting lipid-lowering drug therapy for primary prevention of coronary disease: an updated Sheffield table, *Lancet*, 348, 387–8.

Ramsay, L.E., Haq, I.U., Jackson, P.R. and Yeo, W.W. (1996b) Sheffield risk and treatment table for primary prevention of CHD (Authors' reply), *Lancet*, 348, 1040–41.

Reynolds, T.M. *et al.* (1997) Standing Medical Advisory Committee should reconsider advice to use Sheffield risk table, *British Medical Journal*, 315, 1315.

Ritchie, L.D. and Isles, C.G. (1999) Targeting lipid lowering drugs for the primary prevention of coronary heart disease: comparison of Sheffield Table and New Zealand Guidelines, *British Journal of Cardiology*, 6, 220–2.

Royal College of General Practitioners (1992) *Guidelines for the Management of Hyperlipidaemia in General Practice. Towards the Primary Prevention of CHD.* Occasional Papers No. 35. London: RCGP.

Royal College of General Practitioners (1982) *Promoting Prevention.* Occasional Papers No. 27. London: RCGP.

Shaper, A.G., Pocock, S.J., Phillips, A.N. and Walker, M. (1986) Identifying men at high risk of heart attacks: strategy for use in general practice, *British Medical Journal*, 293, 474–9.

Shepherd, J., Cobbe, S.M. and Ford, I. *et al.* (1995) Prevention of coronary heart disease with pravastatin in men with hypercholesterolaemia, *New England Journal of Medicine*, 333, 1301–7.

Truett, J., Cornfield, J. and Kannel, W. (1967) A multivariate analysis of the risk of coronary heart disease in Framingham, *Journal of Chronic Disease*, 20, 511–24.

Tunstall-Pedoe, H. (1991) The Dundee coronary risk-disk for management of change in risk factors, *British Medical Journal*, 303, 744–7.

Vallance, P. and Martin, J. (1997) Drug therapy for coronary heart disease: the Sheffield table (Letter), *Lancet*, 350, 1854.

Wierzbicki, A.S. and Reynolds, T.M. (1996a) Sheffield risk and treatment table for cholesterol lowering in prevention of coronary heart disease (Letter), *Lancet*, 347, 466–7.

Wierzbicki, A.S. and Reynolds, T.M. (1996b) Sheffield risk and treatment table for primary prevention of coronary heart disease (Letter), *Lancet*, 348, 1040–1.

Wierzbicki, A.S. and Reynolds, R.M. (1997) Guidelines need to concentrate on reducing overall cardiovascular risk (Letter), *British Medical Journal*, 315, 1614–5.

Williams, B., Poulter, N.R., Brown, M.J., Davis, M., McInnes, G.T., Potter, J.F., Sever, P.S. and Thom, S.McG. (2004) Guidelines for the management of hypertension: report of the fourth working party of the British Hypertension Society, 2004, *Journal of Human Hypertension*, 18, 139–85.

World Health Organisation (1982) *Prevention of Coronary Heart Disease. Report of WHO Expert Committee.* Team Report Series No. 678. Geneva: World Health Organisation.

Wood, D., Durrington, P., Poulter, N., McInnes, G., Rees, A. and Wray, R. (1998) Joint British recommendations on prevention of coronary heart disease in clinical practice, *Heart*, 80 (Supplement 2), S1–S29.

Chapter 7

Telephone triage, expert systems and clinical expertise
D. Greatbatch, G. Hanlon, J. Goode, A. O'Caithain, T. Strangleman and D. Luff

Introduction

In recent years there has been a growing use of protocols that are designed to standardise medical work procedures. This growing protocol activity is increasing the bureaucratisation, regulation and control of healthcare practices, especially in the work of 'soft' healthcare professionals such as general practitioners and nurses. One of the ways in which such protocols are being embedded into everyday medical practice is through expert computer systems. Like expert systems in other contexts, these technologies are characteristically built on abstract universalised rules, which supposedly capture the knowledge and practices required to perform tasks and/or to resolve problems. Critics of rule-based expert systems, however, argue that they inevitably fail to capture the 'knowledegeability' of users, and can never cover the range of contingencies that arise on a day-to-day basis in work settings (*e.g.* Suchman 1987, Whalen 1996, Whalen and Vinkhuyzen 2000, Heath and Luff 2000, Gorman 2002). In the context of medical work, this argument has been powerfully articulated by Berg (1997a, 1997b, 1999), who suggests that, regardless of whether they are embedded in technology or not, protocols, guidelines and standards for medical work are fundamentally reductionist because they seek the 'single answer' to what are often complex, multifaceted issues. Hence, protocols impose a formally rational, individualist structure on work that in situ is actually social, affective, hermeneutic and collectivist. Whereas there is an organisational desire to standardise and achieve consistency, for Berg what should be celebrated is the variation of outcomes. These variations reflect individualised service rather than the limitations of an individual professional's cognitive abilities or the failings often attributed to these professionals (Berg 1997a: 1083).

In examining these issues, this chapter considers how an expert system developed for the British telephone triage service NHS Direct is used by staff in their daily work. NHS Direct is a national 24-hour nurse-led service which offers health advice and information via the telephone. The service – which borrows from models of telephone triage familiar in the USA, Denmark, Sweden and UK GP out-of-hours and Accident and Emergency (A&E) departments (Lattimer *et al.* 1998) – is designed to provide citizens with easy access to professional medical advice on the most appropriate forms of care. It is also supposed to stimulate self-learning and encourage people to take greater responsibility for their own health, thereby lessening the burden on other parts of the NHS (Department of Health 1997). The service has grown rapidly since it was launched in 1997 and received some 3.5 million calls during 2001–2, a figure which doubled in 2003 (National Audit Office 2002). Currently, NHS Direct is the largest telephony healthcare line in the world.

NHS Direct uses an integrated telephony and computerised clinical assessment system (CAS), which is designed to support nurse triage by providing expert clinical reasoning. Senior management make no secret of the fact that they are attempting to use this system to limit the extent to which the conduct and outcomes of nurse triage depends upon the professional clinical judgements of individual nurses (Gann 2002). For example, the introduction to a CAS training manual informs NHS Direct staff that:

CAS ensures a uniform approach to processing a call. This approach minimizes malpractice risk as well as improving call centre performance. The decision support software is owned by the NHS and managed by AXA. Working collaboratively, both organisations strive to improve the software, thereby enhancing and standardising the delivery of nursing care nationally.

Thus, NHS Direct's senior management believe that CAS can be used to limit the autonomy of individual nurses in order to ensure consistency and safety, to lessen risk, and to provide a standard level of health advice regardless of the time of calls, the locality of callers, the sites answering calls, or the different backgrounds and specialties of nurses. However, as we have noted elsewhere (Hanlon *et al.* forthcoming), this strategy clashes with the occupational culture of nurses, which places a premium on individualised, holistic care. This situation creates a tension within NHS Direct triage, as well as within NHS Direct as a whole.

In this chapter we examine how nurses actually utilise CAS and their own knowledge, experience and clinical expertise when triaging calls. Using the methods of conversation analysis (CA), we focus on the 'dispositional phase' of the calls, during which nurses advise callers of the recommended level of care for their symptoms, and provide instructions and advice appropriate to the disposition. We show how the nurses privilege their own knowledge and expertise, and provide an individualised service by adapting, supplementing, overriding/underriding and pre-empting CAS recommendations. We conclude by arguing that NHS Direct management's pursuit of standardisation will achieve only limited success because of the professional ideology of nursing and the fact that rule-based expert systems capture only part of what 'experts' do. Whereas previous studies of the use of medical expert systems have sought to gauge levels of user satisfaction with such systems and/or to establish whether or not they improve clinical decision making (*e.g.*

Gardner and Lundsgaarde 1994, Miller, Pople and Myers 1984, Moore 1994, Zielstorff *et al.* 1997), our research sheds light on an issue that has attracted little systematic analytic attention – the ways in which expert systems feature in the moment-by-moment performance of medical work (Greatbatch, Murphy and Dingwall 2001).

NHS Direct

Callers access the NHS Direct service by dialing a single national number, for which they pay at local rates. Currently, NHS Direct also takes large numbers of calls made to GPs outside surgery hours ('out-of-hours' calls). It acts as a message-handling service for GP co-ops aligned to the service, triaging their calls. Callers whose GPs belong to one of these co-ops are either automatically diverted to NHS Direct when they telephone their doctor's surgery out of hours, or they hear a recorded message giving the number of NHS Direct. The service has set a target of integrating all GP out-of-hours calls by 2004, which will increase call intake to 10 million.

On ringing the national NHS Direct number callers are directed to a NHS Direct site where the call is answered by non-nursing call handlers. At periods of high demand, staff may divert calls to less busy sites in other parts of the country but generally calls are put through to a local site. The call handlers get the biographical details of the callers and make initial decisions about the urgency of the calls, using a part of CAS that is specifically designed for use by them. The call handlers are able to connect callers directly to the ambulance service where necessary. Call handlers assign a numerical grade to each call before putting it in the queue for nurse triage. How quickly nurses ring back calls concerning medical problems is dependent on the volume of calls that day and how the call has been ranked by the call handlers in terms of its urgency and type. Those calls which involve requests for health information rather than advice on symptoms are assigned the lowest grade and dealt with by non-nursing health information advisors.

The information from calls that are selected for nurse triage is electronically passed to a nurse by the call handler. Nurses are assisted in their advice-giving by CAS, which is designed around sets of algorithmic questions called protocols. The algorithms are organised in terms of symptoms (such as 'dizziness', 'cough', 'chest pain', 'headache') as opposed to 'conditions' (such as 'diabetes', 'angina', 'migraine'), although there are a few exceptions to this (for example, there is a protocol entitled Chicken Pox). Nurses are expected to establish the nature of the patient's symptoms, enter details of the patient's past medical history (medical problems, current medications and allergies), select an appropriate algorithm, and then ask the symptom-based questions that CAS prescribes. The questions take two forms: (i) those with a 'yes', 'no' or 'uncertain' answer and (ii) those which comprise a list of symptoms. In the case of the former, the nurses click the relevant answer, whereas in the case of the latter, they either click the square box next to the symptom that the patient has or click 'No'.

Three other fields are displayed on the screen alongside the one which contains the algorithmic questions: a box in which nurses can add notes relating to the questions, the patients' symptoms or other relevant information; a box containing the clinical reasoning behind each of the questions; and a box containing details of conditions or medications which the nurse should take into account during the triage process. When the algorithm is completed, CAS recommends a final disposition and issues instructions appropriate to that disposition. Final dispositions include: A&E, immediate or routine contact with GP and home care. Nurses are able to select a higher or lower disposition ('overriding' and 'underriding') than that recommended by CAS as long as they document their reasons for doing so. The need for nurses to document why the software's recommendation was not followed explicitly indicates the presumption that the abstract, universal advice of the technology is more trustworthy than the expertise of the nurse. After a disposition has been selected, the nurse then selects items from a list of care topics in order to summon the advice recommended by CAS. If the nurse gives additional advice that is not recommended by CAS, they are instructed to type this in the 'Advice Recommended' box.

Safety is encoded in CAS in three ways. First, it lessens the risk of callers' conditions worsening before they are assessed, by increasing the speed at which calls are processed. Second, the software starts with the 'worst case scenario' for a symptom in order to eliminate or deal appropriately with emergencies before proceeding with the assessment. Finally, it minimises the risk of malpractice by providing nurses with sets of logically structured symptom-based questions, which are underpinned with 'evidence-based' rationales. The system is designed to offer a minimum safe standard of assessment and advice regardless of the different nursing backgrounds and specialties of the nurses operating it, and a consistent service regardless of the locality of the caller or the site answering the call. In view of this, NHSD's senior management, in its efforts to deliver safety and consistency, are attempting to limit nurses' reliance on their own professional clinical judgement in preference to the universal expertise embodied in CAS (Gann 2002).

Data and method

The research discussed in this chapter was conducted as part of a wider project on NHS Direct entitled NHS Direct: Patient Empowerment or Dependency, which was funded by the ESRC/MRC Innovative Health Technologies Programme (grant number L218252022). The project collected and analysed a variety of data including: field notes based on participant and non-participant observation; in-depth semi-structured interviews with health professionals and callers; and audio recordings of calls to two NHS Direct sites. The present chapter is based on the analysis of 60 recordings of calls to one of the sites, together with copies of the CAS call reports.

The analysis of the calls draws on the approach and findings of conversation analysis (CA), which involves detailed qualitative analysis of audio and video recordings of naturally occurring social interactions (Atkinson and Heritage 1984, Heritage 1995, Psthas 1995). CA research does not entail the formulation and empirical testing of a priori hypotheses. Rather, it uses inductive search procedures to identify regularities in verbal and/or nonverbal interaction. The objective is to describe the practices and reasoning that speakers use in producing their own behaviour and interpreting the behavior of others. Analysis emerges from the orientations and understandings that parties unavoidably display to each other during their interactions.

In locating and analysing recurring patterns of action and interaction, CA researchers repeatedly replay audio or video recordings of natural interactions, carefully transcribing the events. The transcripts capture not only what was said, but also various details of speech production, such as overlapping talk, pauses within and between utterances, stress, pitch and volume. They may also track visual conduct such as gestures and gaze direction. These transcripts facilitate the fine-grained analysis of recordings, enabling researchers to reveal and analyse tacit, 'seen but unnoticed' (Garfinkel 1967) aspects of human conduct that otherwise would be unavailable for systematic study.

Although CA began from the study of ordinary conversations, it has been applied increasingly to other forms of interaction, including medical consultations, broadcast interviews, calls for emergency assistance, organisational meetings, proceedings in small claims courts and psychiatric intake interviews (*e.g.* Boden and Zimmerman 1991, Boden 1994, Drew and Heritage 1992, Maynard and Heritage 2006). A number of researchers have also extended its principles to the study of visual conduct (*e.g.* Heath 1986, Goodwin 1981, Heath and Luff 2000), as well as to the use of computer systems in professional/client interactions (*e.g.* Whalen 1996, Whalen and Vinkhuyzen 2000, Greatbatch 2006, Greatbatch *et al.* 1995, Greatbatch *et al.* 2001). Despite its name, CA is a generic approach to the study of social interaction.

Findings

In this section, we focus on the ways in which nurses present, override/underride and pre-empt recommendations made by CAS.

Reporting CAS dispositions and advice
When nurses follow the recommendations of CAS, they rarely indicate to callers that they are doing so. Consider the following example, which is drawn from a case concerning a 14-year-old schoolboy who has been kicked twice in the testicles by a fellow pupil. (The transcripts included in this

chapter have been anonymised through the use of fictional names of people and places.)

Extract 1 [Testicular Swelling]
```
 1 Nurse:        Is it burning when you're passing uri:ne. [Does it hurt.
 2 Son:                                                    [No.
 3 Son:          (No).
 4 Nurse:        No.
 5               (.)
 6 Nurse:        Okay.
 7               (.)
 8 Nurse: -->    What I'm going to do I'm going to get you to ( ) be seen by
                 the doctor Paul.
 9 Son: -->      Yeah?
10 Nurse:        Okay.=Is your mum there.
11               (.)
12 Son: -->      Yeah. ((Addresses mother)) I'm going to get seen by a doctor.
13               (.)
14 Mother:       Hello duck.
15 Nurse: -->    Hello there. I'm going to arrange for him to see the
                 doctor.
16 Mother: -->   Yeah.=
17 Nurse:        =D'you know where the uhm med- medical centre is on
                 Dove Road [( )
18 Mother:                 [Yeah. Well
19               we use it.=They send a mini-bus out for us because
                 they. . . .
```

[Extract from the Call Report]
   Disposition:
   Contact GP Practice within 4 Hours (as soon as possible)
   Advice Recommended:
   The symptoms you have described during this call suggest that you (or the person concerned) should be assessed by your GP as soon as possible (at least within 4 hours).

After the nurse has entered the boy's response to the last of the set of algorithmic questions prescribed by a protocol entitled 'Testicular Swelling' (lines 1–4), CAS recommends contact with a GP within four hours. The nurse accepts this disposition and the CAS advice that is based upon it. However, she presents the disposition and advice as her own ('I'm going to get you to be seen by the doctor') (lines 8–16). In this way she presents herself to the boy and his mother as an independent professional, bringing her own knowledge and expertise to bear, rather than as, for example, a cipher or a 'go-between' for a computerised system.

In addition to refraining from attributing dispositions and advice to CAS, the nurses also tailor CAS recommendations to constraints and contingencies associated with particular cases, patients and callers – for example by reordering advice/information and/or qualifying recommendations. Consider the following case, in which a woman has contacted NHS Direct about a chest infection, for which a doctor has prescribed antibiotics. The woman has called NHS Direct to find out 'what cough medicine she can take for relief at the moment' (Comments section in Call Report). In this instance the nurse formulates the CAS disposition as the NHSD recommendation (note the use of 'we' rather than 'I'), but characteristically she does not mention CAS (line 53).

Extract 2 [Cough]
```
 1 Caller:      Just to check [yeah.
 2 Nurse:                     [Are you feeling feverish?
 3               (.)
 4 Caller:      er (.) A little bit but not much. I've got more of a headache
               really (than) (.) I mean
 5              if (.) I wasn't on these tablets I'd take a paracetamol and
               then it'd proba[bly go=
 6 Nurse:                      [hm hm
 7 Caller:      =you know.
 8              (.)
 9 Caller:      ( ) coughing [( )-
10 Nurse:                    [Have you got any pain when you're coughing.
11              (.)
12 Nurse:       In your chest?
13 Caller:      No:.
14 Nurse:       (N[o:)
15 Caller:         [My lung hurts a bit. my right lu:ng. (.) At the back. You
               know when I cough.
16 Nurse:       Ri:ght.=
17 Caller:      =That hurts a bit.
18              (.)
19 Caller:      I've been hoping these tablets would clear it all up you see.=
20 Nurse:       Ye::s.
21              (.)
22 Nurse: -->   Are you drinking plenty of warm liquids.
23 Caller:      Ye:s.=Oh yes.
24              [SEVERAL LINES OMITTED ]
25 Nurse:       So if your sister's going out today::
26 Caller:      ( )
27 Nurse:       she could pop into the chemist
28 Caller:      Ye:s
29 Nurse:       speak to a pharmacist
30 Caller:      Yes
```

31	Nurse:	just to make sure (.) that you're- you're able to take the Benylin.
32		(.)
33	Nurse:	But I think (.) fro- from what they- the tablets that you've told me that you're already o:n
34	Caller:	Mhm:
35	Nurse:	you are all right to take the [Benylin as well.
36	Caller:	[(The Benylin)
37	Caller:	Yes. But to make sure.
38	Nurse:	Just to make sure.
39	Caller:	mhm
40	Nurse:	If she just pops into the chemist
41	Caller:	Yeah
42	Nurse:	just speaks [to the pharm-
43	Caller:	[Well we're only a block away from Safeway you see.
44	Nurse:	Oh right you're right near Safe[way are you you.
45	Caller:	[O : : h y e s and she regularly goes [there.
46	Nurse:	[Yes.
47	Nurse:	And just to be on the safe side because they might say oh yes it's all right but she
48		might be better trying (this).
49	Caller:	( ) yeah.=
50	Nurse:	=uh:m (.) You know that's [(wha)-
51	Caller:	[Right oh
52	Nurse: -->	That's what we would recomme[nd for you.
53	Caller:	[Yes ( )-
54	Caller:	Okay ( )
55	Nurse: -->	Now if [things
56	Caller:	[(Thanks very much)
57	Nurse: -->	if things are not settling do:wn
58	Caller:	mhm:
59	Nurse: -->	you know sort of after the weekend
60	Caller:	Yes
61	Nurse: -->	then I'd pop back to see your doctor.
62	Caller:	mhm
63	Nurse: -->	uhm just to see if- if- if they want to examine your chest again just to see if things are settling down [for you.
64	Caller:	[( ). [(Yes)
65	Nurse: -->	[uhm But obviously
66		if you get any problems between now and then you can [always ring- ring back.
67	Caller:	[(You see- well uh-) he
68		gave me three Wednesday ni:ght
69	Nurse:	Mhm

70 Caller: one I took straight away and then two the next morning (.)
               because I hadn't time to go to the er
71   (.)
72 Nurse: mhm
73 Caller: er the pharmacy
74 Nurse: Yeah.
75 Caller: but- so these tablets (I feel) they probably haven't taken
               hold yet.

[Extract from Call Report]
   Disposition:
   Routine Appointment with GP
   Advice Recommended:
   The symptoms you have described during this call suggest that you (or the
   person concerned) should make a routine (non-urgent) appointment to
   discuss matters further with the GP. If the condition changes or worsens
   in any way or if new symptoms appear or arise, call NHS Direct for a
   reassessment. Warm liquids usually relax the airway and loosen mucus.
   Drink a warm drink with lemon and honey to help soothe a cough.
   Elevate head of the bed to reduce cough at night. Turn down the heat and
   open windows to help suppress cough. Ask the pharmacist about cough
   medicines or other medicines.

At lines 1–32 the nurse's queries address two of the final three questions
in the algorithm entitled 'Cough'. The nurse does not ask the last of the
prescribed algorithmic questions as the patient has already furnished the
information concerned in response to an earlier question. At this point CAS
formulates a disposition (Routine Appointment with a GP) and then, after
the nurse selects the care topic(s) that she deems appropriate to the information gathered during the triage process, populates the 'Advice Recommend'
box with advice. The advice recommended by CAS (see the extract from call
report above) is ordered as follows: (1) a suggestion that the patient should
make a routine appointment with their GP; (2) an assertion that the caller
should contact NHS Direct for a reassessment if the condition worsens, and
(3) a series of measures that the patient can take to alleviate the condition –
drink warm liquids, elevate head, turn down heat, open windows at night, ask
a pharmacist about medicines. The nurse, however, does not convey the advice
in this order. Rather she begins by informing the patient about the measures
that can be taken to alleviate her cough (line 22). It is only after she has
reviewed these measures that the nurse addresses the issue of the patient
seeing a GP and, when she does so, she qualifies the CAS recommendation
by making a routine appointment with a GP conditional on the patient not feeling better over the weekend. The advice to call NHSD for a reassessment if
the condition worsens is conveyed to the patient in an unvarnished fashion.
   The fact that the nurse qualifies the CAS disposition in this case may

explain, at least in part, why she opts to deliver the advice first rather than in the order that is prescribed by CAS. By dealing first with self-care measures, she highlights these measures and underlines her view that a routine appointment with a GP will only be necessary if these measures do not alleviate the patient's condition. In relation to this it is worth noting that the advice-delivery format she uses closely resembles a format that the NHS Direct nurses commonly use to convey Home Care dispositions. Rather than explicitly stating that the patient's condition can be managed at home, they provide advice that *presupposes* that the case can be managed through home care. An example of this practice is observable in Extract 3 below.

Extract 3 [Abdominal Pain – Case 1]
((The nurse has used the algorithm entitled 'Abdominal Pain'))

1	Nurse:	Uhm Have you tried any medication.
2	Caller:	Only ((Name of Drug)) tablets.
3	Nurse:	Right. You- you bought that yourself today didn't you?
4	Caller:	Yeah. And it did- did (.) it did go a bit [and then it came back.
5	Nurse:	[Yeah.
6	Nurse:	What- what have you done with regard to eating and drinking today.
7		(.)
8	Caller:	Just had some soup. I can't take food.
9	Nurse:	Right so you've- you've just had soup. And you're able to drink fluids are you.
10	Caller:	Yeah.
11	Nurse: -->	uhm Have you tried anything like any paracetamol or any other medication like
12		that?
13	Caller: -->	No.
14	Nurse: -->	You haven't. Have you got anything at home?
15		(.)
16	Caller: -->	Just the ((name of tablets))
17	Nurse: -->	Okay. Any paracetamol?
18		(.)
19	Caller: -->	No I haven't actually.
20	Nurse: -->	Anything similar like co-codamol or anything like that.=
21	Caller: -->	=I think I have got co-codamol.
22		(.)
23	Nurse: -->	[Yeah.
24	Caller: -->	[And- what they gave me when I hurt my shoulder.
25	Nurse: -->	Right.
26	Caller: -->	(They'll be out of date now).
27	Nurse: -->	Just check the date they should have- they should have a- a date on it.

28 Caller: -->   Yeah. [Yeah.
29 Nurse: -->   [I- I think it's worth trying something like that at the
           moment.
30   (.)
31 caller: -->   You do?
32 Nurse: -->   Yeah. And seeing you know just seeing how the fa- you
           know whether it settles
33   the pain.
34 Caller: -->   [( )
35 Nurse: -->   [With ch-
36 Caller: -->   Yeah.
37   (.)
38 Nurse: -->   I mean obviously if the pain it becomes increasingly worse
           just call us back. Uhm
39   (.)
40 Caller: You don't think it's constipation do you?
41 Nurse: When did you last have your bowels open?
42 Caller: Well this morning . . .

[Extract from the Call Report]
   Disposition:
   Home Care
   Advice Recommended:
   The symptoms reported during this call suggest that the problems
   concerned can be managed at home. WORSENING: If the condition
   changes in any way, or worsens or if any new symptoms arise call NHS
   Direct back for a reassessment.

After asking the last of the algorithmic questions (lines 1–5), and then
establishing that the caller is able to take fluids (lines 6–10), the nurse
advises the patient as to what she should do (lines 11+). The advice recommended by CAS simply states that the problem can be managed at home
but that the patient should call back if the condition worsens or new
symptoms arise – this may be because the nurse has not selected an item
from the care topics menu. The nurse does not overtly inform the patient
that his condition is suitable for home care. Instead, she suggests that he
should take pain killers, thereby leaving him to infer that the recommendation is home care. In so doing, she first gets him to reconfirm that he has
not taken any pain killers (lines 11–14) and then asks whether he has any
tablets in the house that he could take (lines 14–36). In line with the CAS
recommendation, she then indicates that he should call back if the pain
'becomes increasingly worse' (line 38). Returning to Extract 2, one can see
similar processes, with the nurse providing advice which presupposes the
patient's condition can be self-managed, before (unbeknown to the patient)
qualifying the CAS disposition so that the routine appointment with a GP

becomes contingent on self-care not leading to an improvement in the patient's condition.

In summary, the nurses routinely refrain from explicitly linking dispositions and advice to the CAS system. Leaving aside emergency A&E cases (of which we have only a few cases), the nurses do not invoke CAS to warrant final dispositions or advice, or to reassure callers. During their interactions with callers, they present dispositions and advice as emanating from themselves, through the use of the personal pronoun 'I' (see Extract 1 lines 8 and 15, Extract 2, line 62, and Extract 3, line 29) or NHS Direct, through the use of the personal pronoun 'we' (see Extract 2, line 52), without any mention of CAS. Moreover, as Extract 2 shows, the nurses readily adapt, rearrange, qualify and supplement the 'packages' of advice recommended by CAS. Consequently, even when nurses do not override or underride CAS dispositions, they provide callers with advice which is tailored to their particular circumstances, and which draws on the nurses' knowledge, expertise and experience. Interestingly, although nurses are instructed to type details of any additional advice they give to patients in the Advice Recommended box, the nurses do not always do so. Thus, for example, in Extract 3 above the nurse does not record in CAS that she has advised the patient to use pain killers.

Overriding and underriding CAS dispositions and advice
As noted above, nurses are able to 'override' or 'underride' the dispositions recommended by CAS, although they must document their reasons for doing so. Generally, the nurses *do not* indicate to callers that they are overriding/underriding dispositions recommended by CAS. Once again, then, they background CAS in their discussions with callers by not mentioning it. In the following example a mother has called to seek advice concerning her daughter who is suffering from abdominal pain, and the nurse selects a protocol entitled 'Abdominal Pain, Child (Age 5–16 years)'. CAS recommends Home Care, but the nurse opts to override this and suggests a next-day appointment with a GP. However, she does not indicate to the caller that she is overriding/underriding a recommendation made by the system. The override is thus accomplished covertly insofar as the caller is concerned (lines 11–20).

Extract 4 [Abdominal Pain – Case 2]
 1 Caller: I mean the first I thing I thought of was er: a grumbling
                appendix.
 2 Nurse: Ri[ght yeah:.
 3 Caller: [( )-
 4 Caller: (So I thou[ght)
 5 Nurse: [But children (don't) ( ) is different than adults
                because they can
 6   feel
 7   their pain in different areas.

```
 8 Callers:        This is it. [Yea:h.
 9 Nurse:                      [(And it ( ) to be very really accurate with
                               [pain as well.
10 Caller:                     [I kno:w. I know.
11 Nurse: -->      I would ( ) carry on with paracetamol tonight your hot
                   water bottle (.) we're
12                 here all night and we have got access to the GPs. [If Alison
                   wakes up crying=
13 Caller: -->                                                      [Right.
14 Nurse: -->      =and you're worried about anything that can't wait until
                   morning.=
15 Caller: -->     =Yeah.
16 Nurse: -->      By all means come back to us and we'll get her looked at tonight.
17 Caller: -->     Okay.
18 Nurse: -->      Failing that I would be ringing the surgery tomorrow and
                   getting her looked at by
19                 your own GP.
20 Caller: -->     A::ll right. ( )
```

[Extract from the Call Report]
    Disposition:
    Contact GP Practice within 36 hours (next day appointment)
    Disposition Override From:
    Home Care
    Disposition Override Reason:
    Feel caller's symptoms warrant higher level of care than recommended.
    Advice Recommended:
    The symptoms you have reported during this call suggest that you (or the
    person concerned) should be assessed by the GP within the next 36 hours.
    If the condition changes or worsens in any way or if any new symptom or
    symptoms appear or arise, call NHS Direct for a reassessment. MUM
    ADVISED HOME CARE OF PARACETAMOL, AND COVERED
    HOT WATER BOTTLE TO ABDOMEN QUITE APPROPRIATE.
    ADVISED TO SEE OWN GP TOMORROW IN VIEW OF
    SYMPTOMS PERSISTING. TO CALL BACK TONIGHT IF
    SYMPTOMS INCREASE/WORRIED.

In addition to covertly overriding the CAS disposition, the nurse provides
the caller with advice over and above that which is recommended by CAS in
relation to the disposition she has selected. CAS simply recommends a routine appointment with a GP and an NHS Direct reassessment if the caller's
condition worsens or new symptoms arise. Before conveying this advice to
the caller, however, the nurse advises the mother as to how she should care
for her daughter in the meantime (line 11).

Another example of a nurse overriding/underriding a CAS disposition without informing the caller that they are doing so is observable in the following extract. Here the nurse is speaking to a woman who has called on behalf of her 59-year-old mother, who has had a sudden attack of dizziness and hot flushes. Following the completion of the set of algorithmic questions entitled 'Seizure', the nurse double-checks that the patient is 'conscious. . . . and with it' (lines 1 and 3). Given that the CAS system is recommending that the patient be taken to A&E as soon as possible, and that the nurse subsequently underrides this disposition, the questions seem to be designed to establish whether such an underride is advisable. Having established the whereabouts of the woman and her daughter (lines 3–12), she then underrides the CAS disposition by advising the woman that she should take her mother to a nearby medical centre (lines 13–18).

Extract 6 [Seizure]

```
 1 Nurse:        =Okay and is she quite conscious.
 2 Caller:       Yes,
 3 Nurse:        And quite with it. [Right. And where do you live?
 4 Caller:                          [hm hm
 5               (.)
 6 Caller:       u::hm (.) ( [ )-
 7 Nurse:                    [Yeah. Where are you right now.
 8 Caller:       We're in Hazelwall Avenue.
 9 Nurse:        Is that a- Kingsley.=
10 Caller:       =It is.=
11 Nurse:        =Is that Denford.
12 Caller:       Yeah.=
13 Nurse: -->    =Right. What I'd think I'd like you to do: is to nip down
                 to Denford Medical
14               Centre so that we can have a look at [her for you:.
15 (Caller): -->                                     [Mhm
16 Caller: -->   Mh[m
17 Nurse:   -->    [Is that- is that all ri:ght?
18 Caller: -->   Yes.=
19 Nurse:        =Do you know where we are.=
20 Caller:       In Wallfield Road.
21 Nurse:        That's right. It's in Wallfield Road. .hh If you bear with
                 me I'll see if I can give you an appointment time.
22               =Just hold. .hh I'm- I haven't got anywhere it'll just go er
                 mute for a while.
23 Caller:       Yes.
24 Nurse:        ((on phone to Medical Centre)) Molly can I have an
                 appointment time for a lady to
25               come down to you. (.) She's coming from Kingsley . . .
                 [RECORDING ENDS]
```

[Extract from the Call Report]
   Disposition:
   Contact GP Practice within 4 Hours (as soon as possible)
   Disposition Override From:
   Accident & Emergency as soon as possible
   Disposition Override Reason:
   Caller not comfortable with recommended level of care

In the call report the nurse indicates that she underrides the CAS disposition because the patient was not comfortable with the recommended level of care. Presumably, this means that the patient did not want to go to A&E, but there is no evidence of this in the call. And, as we noted above, the nurse does not indicate to the caller that she is underriding the CAS recommendation. Instead, the nurse straightforwardly advises the caller that she wants her to take her mother to a medical centre (lines 13–14). In this case the nurse subsequently checks that the caller is happy with this disposition (line 17), and the patient minimally confirms that she is (line 18).

Occasionally, nurses do overtly indicate to callers that they are 'overriding' or 'underriding' CAS dispositions. The following example is drawn from a case in which a woman has called NHS Direct to seek advice concerning a 'severe headache'. The transcript begins immediately after the nurse has asked the patient the last of the algorithmic questions prescribed by the CAS protocol entitled 'Headache'.

Extract 6 [Headache]
1	Nurse:	There's loads of over the counter pain relief but it sounds like you've had quite a
2		bit since yesterday already.
3	Caller:	Yeah.
4	Nurse: -->	And it's not moving it. So what I sugge- I mean what it's advising me is for you-
5		for you to go to Accident & Emergency but (.) you know I think you'd better go
6		to see your GP today. You ring them up.
7	Caller: -->	Yeah.
8	Nurse: -->	You tell them that you've been in touch with NHS Direct.
9	Caller: -->	Yeah.
10	Nurse: -->	And that we've advised you that you be seen. Because (.) uhm maybe- I'm just
11		looking at your age may be your blood pressure's gone up for some reason. (.) All
12		right. (.) What about your eyes. When did you last have (.) your eyes tested.
13	Caller:	U:hm probably about a year ago.

[Extract from the Call Report]
   Disposition:
   Contact GP Practice within 4 hours (as soon as possible)
   Disposition Override From:
   999 – Ambulance as soon as possible
   Disposition Override Reason:
   Left blank by the nurse
   Specify:
   Left blank by the nurse
   Advice Recommended:
   The symptoms you have described during this call suggest that you (or the person concerned) should be assessed by your GP as soon as possible (at least within 4 hours). Hang up the phone after this call and contact the GP practice of the person concerned and ask for an emergency appointment with the GP.
   MRS SMITH HAS A SEVERE HEADACHE AND CANNOT TOLERATE THE LIGHT IN HER EYES. SHE IS ALERT THEREFORE I HAVE ADVISED THAT SHE SEES HER GP TODAY

Having begun to deliver advice to the caller (lines 1–2 and 4), the nurse cuts off ('So what I sugge-') and informs the caller that 'it' (CAS) is recommending that the caller should go to A&E but that she believes it would be better for the caller to have an emergency appointment with a GP (lines 4–6). In this case, then, the nurse overtly underrides the disposition recommended by CAS and informs the caller that she is doing so. However, although she enters her reason for underriding the CAS disposition as a note in CAS (rather than in the slot provided by CAS) – SHE IS ALERT THEREFORE I HAVE ADVISED THAT SHE SEES HER GP TODAY, she does not share this information with the patient. Cases in which nurses overtly override/underride CAS dispositions appear to be triggered by the arrival of 'unexpected' dispositions on the screen and/or the perceived need to give careful consideration to 'expected' dispositions due to the details of particular cases. In this case, for example, the CAS recommendation (A&E) is at odds with the nurse's approach thus far in-the-call, for she has been speaking about over-the-counter pain relief, etc.

   Regardless of whether they covertly or overtly override/underride CAS recommendations, the nurses privilege their own knowledge and expertise, and present dispositions to patients as emanating from them rather than CAS. This is consistent with their conduct when they convey CAS dispositions to callers, for once again it involves them orienting to the technology as a tool, which is subservient to their own knowledge and experience. The prevalence of 'covert' overrides may have evolved to deal with the fact that when nurses inform patients that they are overriding/underidding CAS they run the risk of patients becoming anxious or concerned about the level of care that is being recommended by the nurse.

Pre-CAS dispositions

Sometimes nurses may inform callers of dispositions prior to CAS recommendations becoming available. When they do so, they proffer dispositional statements before, during or immediately after working through the sets of algorithmic questions prescribed by CAS. In some cases, the nurse's recommendations are consistent with the dispositions that CAS subsequently presents. Consider Extract 7, in which the nurse articulates a disposition before she has even entered a CAS algorithm. The caller has indicated that pus is oozing from a surgical wound on the side of her neck.

Extract 7 [Wound Infection]
```
 1 Caller:       so that's what I was worried ab[out ( )
 2 Nurse:                                        [Yeah- you- you're quite
                 right er we don't want the
 3               infection to get hold do we. [So
 4 Caller:                                     [( )=
 5 Nurse: -->   =we do- we do need to get you looked at. We need to get
                 you looked at anyway
 6               but that makes it more significant [doesn't it.
 7 Caller:                                          [Yeah.
 8 Caller:       Yes.
 9 Nurse:        ( ) I need to go through a process to get this sorted out. [So
                 bear=
10 Caller:                                                                  [( )
11               with me.
12 Caller:       [( )
13 Nurse:        [.hhhhhh
14               (.) Lengthy Silence
15 Nurse:        Any other medical problems.
16               [SEVERAL LINES OMMITED]
17 Caller:       [Yeah.
18 Nurse:        [bear with me a second and I'll put it down as an (wound)
                 infection.
19               (.)
20 Nurse: -->   A few more questions and then we get- .hhh what we need
                 to do Brenda is your
21               own GP being covered by the out of hours surgery,
22 Caller:       Ri::ght?
23 Nurse:        ( ) (and) (going to get you an appointment now and get you
                 looked at).
24 Caller:       Right. Okay.
25               (.)
26 Nurse:        Just bear with me a second. (A few more questions)
27               (.)
28 Caller:       Do I have to go down to the surgery then.
```

29 Nurse:	You will. [We've just got a few more questions and then I can (bat it) all the=	
30 Caller:	[Right (okay).	
31 Nurse:	=to- [to the doctor then.	
32 Caller:	[(Yeah).	
33 Caller:	Right. Okay.	

After the nurse states that she thinks the caller's wound is infected, the caller indicates that she is particularly concerned about this because she recently had a kidney transplant; indeed, she presents this concern as the reason for her call ('so that's what I was worried about' – line 1). Subsequently, the nurse confirms that the caller is right to be concerned and then announces that the caller requires medical attention ('We do need to get you looked at . . .'), although the nurse does not indicate at this stage what this will involve (lines 5–6). After establishing what medications the caller is taking and whether she has any allergies (datum not shown), the nurse announces that she is classifying the patient's symptoms as a 'wound infection' (line 18), which is the title of the CAS protocol that she chooses. As she enters the protocol, the nurse indicates to the caller that the outcome of the triage will be an appointment with a GP (lines 20–24). The questions prescribed by CAS are thus presented to the patient as preliminaries to the arrangement of the GP appointment. After the nurse completes the algorithm, the CAS system presents a disposition that is consistent with the one the nurse has formulated:

[Extract from the Call Report]
  Disposition;
  Contact GP Practice within 12 Hours (same day)
  Advice Recommended:
  The symptoms you have described during this call suggest that you (or the person concerned) should be assessed by a GP within the next 12 hours. If the condition worsens in any way or if any new symptom or symptoms appear or arise, call NHS Direct for a reassessment. PATIENT ADVISED OWN GP BEING COVERED BY OUT OF HOURS PCC, WILL BE CALLED BACK WITH APPT TIME TO ATTEND TODAY.

As the notes that she appends to CAS's advice indicate, the nurse arranges for a local medical centre to call the patient to arrange an appointment because her own GP is being covered by out of hours PCC:

Extract 7 (Continued)
34 Nurse: No problem. We've got all the details we need from you now Brenda. So what I
35   can do if you're happy to do that I can pass all this information to the medical
36   centre in Denford. .h[h

37 Caller: [Ri:g[ht.
38 Nurse: [They   will   call   you   back
         Brenda,=they'll ring you back with an
39   appointment a time for you to attend this afternoo:n. .hh[hh
40 Caller: [Yeah
41   [but ( )-( )- your normal doctors.
42 Nurse: [( )
43 Nurse: No no it's actually in Denford,
44 Caller: Oh ri::ght.=
45 Nurse: =They- they can give you some in- some- some directions to get
         there.=It's Field
46   Road in Denford. It isn't difficult to find. Don't worry.=
47 Caller: =All rig[ht.

It is possible that in cases like this nurses demonstrate an ability to anticipate
CAS dispositions, as well as using their own knowledge and expertise to
formulate dispositions independently of CAS. The nurse's anticipation of the
CAS disposition is perhaps based on inferences/knowledge she draws from
the opening segment of the call, combined with her knowledge of the likely
outcome of the algorithm given the symptoms the patient/caller has already
described. Of course, this does not explain why the nurse opts to communicate
the disposition to the caller before working through the CAS algorithm.
With regard to this, it is noticeable that, as in other cases in our data base, the
nurse subverts the system after the caller has expressed anxiety concerning
her condition. Thus her actions may be directed towards alleviating the
caller's anxiety by reassuring her that she will receive an appropriate level
of care.

When the nurses finish pre-CAS dispositions that are not consistent with
those which are subsequently proffered by CAS, they generally covertly
override/underride the system's recommendations. In the following example
a father has called NHS Direct to discuss his daughter, a four-year old who
has chickenpox (diagnosed by a GP). As the nurse asks him the algorithmic
questions in the protocol entitled 'Chickenpox', the caller expresses anxiety
concerning his daughter – first asserting that he's worried that the scabs
under his daughter's arms may be going septic (data not shown), and then
expressing concern about the spots around her eyes (lines 1 and 4). After the
latter expression of anxiety, the nurse proffers a disposition, even though
she has yet to complete the algorithmic questions prescribed by CAS (lines
12+).

Extract 8 [Chickenpox]
1 Caller: She's got a lot of spots around her eyes though.
2 Nurse: Ri:ght. Okay.
3   (.)
4 Caller: -->   I'm getting worried about that as w[ell.

```
 5 Nurse:                                          [Yeah.=
 6 Caller:       =( )
 7 Nurse:        Right. Or: which one's?
 8               (.)
 9 Caller:       ( )
10 Nurse:        ( ) Right. How long is it that she's- that she's had the
                 chickenpox.
11 Caller:       She's had it about three days now. (.) Two or three days.=
12 Nurse: -->    =Right. (.) Obviously (.) what we're going to do: uhm I've
                 got to put it through to
13               the GP.=.hh Now you're obviously going to need to take
                 her down to the medical
14               centre but what I think I'd better do:: .hh is give them a ring
                 fir:st=
15 Caller:       =Yeah.=
16 Nurse: -->    =and find out (.) see if we can get an appointment or
                 something.=because .hh
17               obviously you can't take her down there with other people.
                 =They need to be able
18               to see you straight away. .hh So if you just give me five or
                 10 minutes I'll actually
19               ring them and- and find out what the best thing to do
                 is.=All ri::ght.
20 Caller:       Yea[h.
21 Nurse: -->       [And then- d'you know where the emergency medical
                    center is on that-
22               Wallfield uhm- so- sorry I'm thinking of Denford .hhh o:n
                 er Moor Roa:d?
23 Caller:       u::hm No.
24               (.)
25 Caller:       ( )
26 Nurse:        Right. Er Let me just look and see ( )- you said- you said
                 she hasn't got
27               cough.=She's not bringing any mucus or anything up.
28 Caller:       No.
29 Nurse:        No.=Okay. Ri:ght.=Is tha- you know that eyes's that's-
                 that's shocking is that
30               discharging at a:ll.
31 Caller:       It's er like [( )=
32 Nurse:                     [watery
33 Nurse:        ah: oh She sounds really poorly to be honest.
34 Caller:       Yeah.
35               (.)
36 Nurse:        Oka::y. U:::hm (.) Bear with me I'm just going to have a
                 look ( ). (Here
```

37		we go).=
38	Caller:	=( ).
39	Nurse:	u:::hm Well I think that's the other thing that it may that because she's got
40		chickenpox we may be able to get her in at- .hh what I'll do Paul is ring the
41		emergency centre first.
42	Caller:	Yeah=
43	Nurse:	=And find out. If you have to go down there:
44	Caller:	(I [ )
45	Nurse:	[Well I mean it depends on what the doctors feel
46		about it,
47	(Caller):	( )
48		(.)
49	Nurse:	I think what we'll do I'll put it through for advice.=Because they can- the doctors
50		can talk to you directly
51	Caller:	m[hm
52	Nurse:	[and then you and they will be able to make the decis[ion. I mean in the=
53	Caller:	[Right.
54	Nurse:	=circumstances I would have hoped that they might come out and see her.

[Extract 1 from Call Report]
  CHILD HAS HAD CHICKENPOX FOR ABOUT TWO DAYS
  CHILD HAS BEEN DRINKING REASONABLY WELL BUT
  TONIGHT SHE IS UNABLE TO FEED HERSELF AS UNDER
  KNEES, IN HER ELBOWS, UNDER ARMS IS YELLOW AND
  STARTING WEEPING AND SHE IS UNABLE TO BEND JOINTS.
  DAD THINKS SHE HAS SCRATCHED HERSELF.
  ALSO HAS SPOTS AROUND HER EYE AND LEFT ONE IS HALF
  SHUT.
  GP ADVICE PLEASE
  DR JOHNSON

[Extract 2 from Call Report]
  Disposition:
  Home Care
  Advice Recommended:
  Left blank by CAS

After conveying the disposition to the caller, the nurse completes the remaining algorithmic questions prescribed by CAS (lines 26–31). After the caller's response to the last of the algorithmic questions, the nurse produces a sympathetically intoned 'ah': and then produces an assessment of the child's condition ('oh she sounds really poorly to be honest') (line 33). She then overrides the CAS system disposition, from Home Care to GP Emergency Appointment, although she does not indicate that she is doing so to the patient. As she develops her disposition, the nurse raises the possibility that a GP may visit the child's home due to the contagious nature of her complaint, and reports that she will arrange for a doctor to telephone the father with respect to this issue (lines 39–54).

In one or two instances, the nurses *do* explicitly indicate to patients that they are overriding/underriding a disposition. As in the cases of overriding/underriding discussed in the previous section, in these cases the nurse appears to be surprised by the level of care being recommended by CAS. Consider the following example in which the caller has been coughing up green phlegm, sweating, feeling dizzy and experiencing tightness in her chest.

Extract 9 [Breathing Trouble]

1 Nurse:		Are you (.) are you finding that you're short of breath or catching your breath.
2 Caller:		Yeah and I've go- uhm (.) and when my chest tightens I start to feel a bit (.) dizzy.
3		(.)
4 Nurse:		And that's with your tightness.
5 Caller:		Yeah.
6 Nurse:	-->	Well obviously you need to be seen by a doctor.
7 Caller:		Yes. Okay.
8 Nurse:		Just hold on.
9		(.)
10 Nurse:	-->	I mean it's saying ambulance but (.) you're breathing okay, you're talking to
11		me [okay.
12 Caller:	-->	[Yeah.
13 Nurse:	-->	I'll get you to see the doctor.
14 Caller:	-->	Okay.
15		(.)
16 Nurse:	-->	Okay.
17		(.)
18 Caller:	-->	((Addresses someone other than the nurse)) I'm going to see the doctor.
19 Nurse:		Are you coughing up any phlegm?
20 Caller:		I've been bringing some up [yeah.
21 Nurse:		[What colour?
22 Caller:		It's like a greeny colour.
23 Nurse:		All ri::ght.

[Extract from the Call Report]
  Disposition:
  Contact GP Practice within 4 hours (as soon as possible)
  Disposition Override From:
  999 – Ambulance as soon as possible
  Disposition Override Reason:
  Left blank by the nurse
  Specify:
  Left blank by the nurse

As the nurse waits for the CAS system to process the responses to the algorithmic questions prescribed by the protocol entitled 'Breathing Trouble', the nurse asserts that the patient will 'need to be seen by a doctor' (line 6) – presenting the disposition as self-evidently the case ('Well obviously'). Note, however, that 'seeing a doctor' could mean the patient being taken to A&E by ambulance, making her own way to a medical centre, or being visited by a GP. The nurse appears to refrain from being more specific prior to the appearance of the CAS disposition on the screen. Subsequently, CAS recommends '999 – Ambulance as soon as possible'. The nurse informs the patient that it has done so, but conveys to the patient that she is minded to underride this disposition. In so doing, she indicates to the patient the reasoning which underpins her decision – the patient is breathing satisfactorily and is talking to her (the nurse) (lines 10–11). Then, after the caller's confirmatory 'Okay' ('Yeah' – line 12), which allows the nurse's characterisation of the caller's condition to stand unchallenged, she overtly overrides the CAS disposition, asserting that she will arrange an appointment with 'the doctor' (line 13). The patient then displays to the nurse that she accepts this ('Okay' – line 14) before indicating to someone who is with her that she is going to see the doctor (line 18). Subsequently, the nurse elicits additional information concerning the patient's symptoms which she records in CAS's free text box before dispatching the details to a medical centre (lines 19+).

In summary, nurses deliver pre-CAS dispositions before, during and immediately after working through the algorithmic questions. In some cases their dispositions mirror those that CAS subsequently recommends, and may in fact involve nurses anticipating the latter on the basis of their past experience (as in Extract 7). In others, the nurses' dispositions differ from CAS recommendations, with the result that the nurses must decide whether to override/underride the latter. In our data base they invariably opt to override/underride CAS, and in so doing generally refrain from informing the callers (as in Extracts 7 and 8, as opposed to Extract 9). In most cases pre-CAS dispositions occur after expressions of anxiety by callers and thus may be directed to reassuring them (as in Extracts 7 and 8). When the nurses present dispositional statements before completing the sets of algorithmic questions, they may treat these questions as exercises in gathering information required to process a case for which a disposition has already been determined. When they do this, CAS's 'dispositional' function is perhaps rendered redundant.

Discussion

CAS is designed to standardise and control nurse-caller interaction. NHSD's senior management want CAS to deliver standardised care to the generalised other and they want it to predominate over the expertise and experience of nurses (Gann 2002). However, our analysis of how NHS Direct nurses use and orient towards CAS dispositions and advice indicates that CAS is not delivering the control that management is looking for. The nurses use CAS in a range of ways and, in so doing, privilege their own knowledge and expertise, and deliver an individualised service. They present CAS dispositions and advice as emanating from themselves either as individuals (through the use of personal pronoun 'I') or as representatives of NHS Direct (through the use of the collective pronoun 'we'), without any mention of CAS. They also adapt, tailor, qualify and supplement the dispositions and advice recommended by CAS. Moreover, when they override/underride CAS recommendations they may or may not opt to indicate to patients why they are doing so. In some cases they even convey dispositions to callers before CAS has made its recommendations.

It is worth noting here that our research indicates that these are not the only ways in which the nurses depart from the recommendations and routines prescribed by CAS. Thus, for example, apart from rephrasing algorithmic questions to ensure callers understand what is being asked, nurses are expected to adopt a uniform approach when using the algorithms. In practice, however, they re-order, conflate, decline to ask and supplement CAS's algorithmic questions (Hanlon *et al.* 2005, Greatbatch *et al.* in prep.). Moreover, they sometimes initiate their own lines of symptom-based questioning independently of CAS, for example after CAS has recommended a disposition (Greatbatch *et al.* in prep.). Or they explore candidate diagnoses offered by patients, which may or may not be consistent with the dispositions and advice that CAS recommends (Greatbatch *et al.* in prep.).

NHSD management's response to the difficulties for NHSD in delivering a standardised service is to suggest that they will be alleviated, if not resolved, by insisting on nurses using the system in a uniform fashion, and ensuring that upgrades of CAS remedy the 'deficiencies' which lead to variations in its use. There are, however, at least two reasons why this approach will fail to fully achieve management's objective. First, as noted above, the concept of standardisation is contrary to the professional ideology of nursing, which emphasises individualised, holistic care (Hanlon *et al.* 2005). The nurse's use of CAS involves them acting in accordance with this ideology. On the one hand, they retain a large measure of autonomy; on the other they present themselves to callers as independent professionals providing an individualised service. With regard to this, it is noteworthy that a recent interview-based study of the use of CAS by NHS Direct nurses in 12 sites found that the nurses continued to value their own expertise and believed that the software had limitations (O'Cathain *et al.* 2004).

Second, the management's strategy is based on the mistaken assumption that the clinical and medical expertise required to ensure consistent and safe assessment can be embodied in abstract, universal rules that operate independently of the characteristics and contingencies associated with particular cases. This assumption is wrong because, as numerous studies have demonstrated, formal rules inevitably have to be adapted, qualified and even disregarded in the context of real world situations (*e.g.* Garfinkel 1967). In the case of NHS Direct this involves nurses tailoring their use of CAS to accommodate a number of human and organisational factors, including the precise nature of a patient's symptoms, whether patients have already received medical assistance (and, if so, what this involved), patients' levels of knowledge about their symptoms, patients' familial and social circumstances, patients' levels of anxiety (as expressed in the calls), the nurses' own and their colleagues' specialist knowledge and expertise, and the nurses' knowledge of how other local NHS services operate in practice.

In addition there are aspects of the professional expertise and reasoning of nurses that resist being transformed into rules that can be embodied in so-called experts systems like CAS. For example, in triaging cases, nurses make a range of judgements, many of them tacit, about the callers and their circumstances. Is the caller exaggerating and/or correctly depicting the symptoms? What does the caller want – reassurance, advice, access to other parts of the NHS or emotional support? Does the caller understand what they are being told? In making these judgements nurses rely on practices and procedures that are largely tacit, and which are rarely discussed or even thought about. These kinds of judgements lie beyond the purview of expert systems.

In conclusion, CAS is viewed by management as an expert system that embodies abstract universalised rules, which provide a basis for uniformity and standardisation of service delivery. However, the nurses use it in ways which privilege their own expertise and enable them to deliver a service which is adapted to the circumstances and needs of individual callers. In part, this reflects the nurses' perceptions of professionalism and concepts of 'holistic nursing'. It is not, however, simply a matter of choice as to whether CAS should strictly control the nurse/caller interaction, or should be used by nurses as a tool, which supports their preferred ways of working. For no matter how sophisticated the CAS algorithms become, they will never cover the vast range of contingencies that confront nurses as they deal with particular cases, or replicate the tacit practices and knowledge that experienced nurses use and rely upon to interpret the conduct of patients/callers. This raises important issues for the development and use of expert systems not only in NHS Direct but also in healthcare organisations in general.

One final point. The ways in which advice and information is delivered in NHSD calls does not solely rest on the conduct of nurses, including the ways in which they use and interact with CAS. NHSD and its personnel do not simply produce a service *for* the consumer, as the producer-consumer interaction is also shaped by the response of the consumer/patient. Like other services (Whalen *et al.* 1988, Zimmerman 1992), the NHSD service is jointly produced by professionals and consumers. Interestingly, NHSD callers rarely question the dispositions or advice that nurses deliver. Even when nurses provide them with clear opportunities to comment on their dispositions, callers adopt a relatively passive stance. In Extract 9, for example, the nurse indicates that she is minded to underride a CAS disposition and indicates her reasons for doing this. Although the patient is thereby given an opportunity to question the decision, she does not do so, thus leaving the way clear for the nurse to follow the course of action she has projected. In Extract 5 the nurse first informs the patient of a disposition and then asks whether this is acceptable ('Is that all right?' – line 17) After the patient responds affirmatively (line 18), the nurse moves on to deal with arrangement making. The fact that callers generally accept dispositions and advice with little or no comment is especially interesting in the light of recent theories concerning the character of professional/client relationships. Giddens (1994) has argued that we inhabit a post-traditional society wherein factors such as globalisation and education alter the relationship between professionals and clients. In short, clients need expert advice (and assistance) more than ever but, increasingly, they reflexively question the advice (and assistance), thereby restructuring the nature of authority within the professional-client relationship as individuals continuously question and challenge the nature of service provision. While it may well be the case that distrust of expertise is growing and people are increasingly prone to criticising professionals, our analysis indicates that these trends have yet to result in significant changes in the ways in which people actually interact with NHSD nurses.

References

Atkinson, J.M. and Heritage, J. (eds) (1984) *Structures of Social Action: Studies in Conversation Analysis*. Cambridge: Cambridge University Press.

Berg, M. (1997a) *Rationalizing Medical Work. Decision Support Techniques and Medical Practices*. Cambridge: MIT Press.

Berg, M. (1997b) Problems and promises of the protocol, *Social Science and Medicine*, 44, 8, 1081–8.

Berg, M. (1999) Patient care information systems and health care work: a sociotechnical approach, *International Journal of Medical Informatics*, 55, 87–101.

Boden, D. (1994) *The Business of Talk: Organizations in Action*. Oxford and Cambridge, MA: Polity Press.

Boden, D. and Zimmerman, D. (eds) (1991) *Talk and Social Structure: Studies in Ethnomethodology and Conversation Analysis*. Cambridge: Polity Press.

Department of Health (1997) *The New NHS*. London: Department of Health.

Drew, P. and Heritage. J. (eds) (1992) *Talk at Work: Interaction in Institutional Settings*. Cambridge: Cambridge University Press.

Gann, B. (2002) The policy perspective, paper presented at the conference on *Information for Patients and Public: the Role of ICTS*. University of Brighton, October 2002.

Gardner, R.M. and Lundsgaarde, H.P. (1994) Evaluation of user acceptance of a clinical expert system, *Journal of American Medical Informatics Association*, 6, 428–38.

Garfinkel, H. (1967) *Studies in Ethnomethodology*. Cambridge: Polity Press.

Giddens, A. (1994). *Modernity and Self-Identity*. Cambridge: Polity Press.

Goodwin, C. (1981) *Conversational Organization: Interaction between Speakers and Hearers*. London: Academic Press.

Gorman, M. (2002) Turning students into professionals: types of knowledge and ABET Engineering Criteria, *Journal of Engineering Education*, July 2002: 327–32.

Greatbatch, D. (2006) Prescriptions and prescribing: co-ordinating talk and text-based activities. In Maynard, D. and Heritage, J. (eds) *Communication in Medical Care: Interaction between Primary Care Physicians and Patients*. Cambridge: Cambridge University Press.

Greatbatch, D., Heath, C.C., Campion, P. and Luff, P. (1995) Conversation analysis: human-computer interaction and the general practice consultation. In Monk, A. and Gilbert, N. (eds) *Perspectives on Human-Computer Interaction*. London: Academic Press.

Greatbatch, D., Murphy, E. and Dingwall, R. (2001) Evaluating medical information systems: Ethnomethodological and interactionist approaches, *Health Services Management Research*, 14, 181–91.

Greatbatch, D., Hanlon, G., Goode, J., Strangleman, T., O'Cathain, A. and Luff, D. (in prep.) Achieving Shared Understandings: Establishing Trust and Managing Risk in Nurse Telephone Triage.

Hanlon, G., Goode, J., Strangleman, T., Luff, D., O'Cathain, A. and Greatbatch, D. (2005) Knowledge, technology and nursing: the case of NHS Direct, *Human Relations*, 58, 2, 147–71.

Heath, C. (1986) *Body Movement and Speech in Medical Interaction*. Cambridge: Cambridge University Press.

Heath, C. and Luff, P. (2000) *Technology in Action*. Cambridge: Cambridge University Press.

Heritage, J. (1995) Conversation analysis: methodological aspects. In Quasthoff, U.M. (ed.) *Aspects of Oral Communication*. Berlin, Germany and Chicago, IL: De Gruyter.

Lattimer, V., George, S. and Thompson, F. (1998) Safety and effectiveness of nurse telephone consultation in out-of-hours primary care, *British Medical Journal*, 317, 1954–9.

Luff, P., Hindmarsh, J. and Heath, C. (eds) (2000) *Workplace Studies*. Cambridge: Cambridge University Press.

Maynard, D. and Heritage, J. (eds) (2006) *Communication in Medical Care: Interaction between Primary Care Physicians and Patients*. Cambridge: Cambridge University Press.

Miller, R.A., Pople, H.E. and Myers, J.D. (1984) INTERNIST-I, an experimental computer-based diagnosis consultant for general internal medicine. In Clancy, W.J.

and Shortcliffe, E.H. (eds) *Readings in Medical Artificial Intelligence*. Reading, MA: Addison-Wesley.

Moore, L.A. (1994) Reaction of medical personnel to a medical expert system for stroke. In Anderson, J.G., Aydin, C.E.A. and Jay, S.J. (eds) *Evaluating Health Care Information Systems: Methods and Applications*. Thousand Oaks, London, New Delhi: Sage.

National Audit Office (2002) *NHS Direct in England*. London: The Stationery Office.

O'Cathain, A., Sampson, F.C., Munro, J.F., Thomas, K.J. and Nicholl, J.P. (2004) Nurses' views of using computerised decision support software in NHS Direct, *Journal of Advanced Nursing*, 43, 3, 280–6.

Psthas, G. (1995) *Conversation Analysis: the Study of Talk-in-Interaction*. Thousand Oaks: Sage.

Suchman (1987) *Plans and Situated Actions*. Cambridge: Cambridge University Press.

Whalen, J. (1996) Expert systems versus systems for experts: computer-aided dispatch as a support system in real world environments. In Thomas, P.J. (ed.) *Social and Interactional Dimensions of Human-Computer Interfaces*. Cambridge: Cambridge University Press.

Whalen, J. and Vinkhuyzen, E. (2000) Expert systems in (inter)action: diagnosing document machine problems over the telephone. In Luff, P., Hindmarsh, J. and Heath, C. (eds) *Workplace Studies*. Cambridge: Cambridge University Press.

Zielstorff, R.D., Estey, G., Vickery, A., Hamilton, G., Fitzmaurice, J.B. and Barnett, G.O. (1997) Evaluation of a decision support system for pressure ulcer prevention and management: preliminary findings. *Paper presented at 1997 American Medical Informatics Association Annual Fall Symposium.*

Zimmerman, D. (1992) The interactional organization of calls for emergency assistance. In Drew, P. and Heritage, J. (eds) *Talk at Work: Interaction in Institutional Settings*. Cambridge: Cambridge University Press.

Chapter 8

Finding dignity in dirty work: the constraints and rewards of low-wage home care labour

Clare L. Stacey

The ageing of the 'baby boom' generation in the US, the UK and other Western countries has spawned a cottage-industry of policy makers, scholars and activists all concerned about the crisis in long-term care for older people (Stone and Weiner 2001). Fuelling much of the concern is the question of who will care for this social group, given that many families no longer live geographically near one another or – if they remain close by – cannot care for a parent or loved one because of pressing demands of career and children (Harrington Meyer 2000).

In the US, these realities are expected to translate into an unprecedented growth in the home care industry. The Bureau of Labor Statistics estimates home care aides belong to one of the fastest-growing occupational groups (Bureau of Labor Statistics 2004). If past conditions hold, those who enter the occupational field of home care will be unskilled, untrained and underpaid (Crown, Ahlburg and MacAdam 1995, Stone and Weiner 2001).

While research to date confirms the demanding aspects of the work and the need for improved working conditions (Crown 1994, Hollander Feldman 1994, Stone 2001, Yamada 2002), little is known about how home care workers themselves experience and negotiate their labour on a daily basis. I address this gap by examining how home care workers navigate a job that is emotionally and physically demanding, stigmatised, and offers few material rewards. In short, given the obvious constraints of the job, how do workers themselves assign meaning to their 'dirty work'?

Qualitative interviews with 33 home care workers (also known as 'aides') and direct observations of home care labour suggest that workers have a conflicted, often contradictory, relationship to their work. Predictably, home care aides felt constrained by the high demands and few material rewards associated with their care. Specifically, workers identified three broad work constraints that compromised their ability to do a good job or to experience their work as meaningful: overwork and added responsibilities; increased risk; and the physical and emotional strain of the job. While these represent significant constraints, workers interviewed also described clear rewards that come from caring for another person. Rewards stemmed from three main sources: practical autonomy on the job, especially relative to prior work in the service sector; skills building; and doing dirty work. I interpret these rewards as key mechanisms through which aides import dignity into a stigmatised and relatively invisible occupation.

Dignity has intrinsic value for the worker, but it also serves to mediate the obvious constraints of poor pay, job insecurity and emotional/physical strain that often accompany paid carework. Although in the long run, the sense of dignity may not prevent worker turnover or burnout, in the short term dignity keeps workers on the job. These findings support recent evidence within the sociology of work and occupations suggesting that workers find ways to manoeuvre and manage risk within the constraints of the new economy (Smith 2001). Given that the healthcare industry generally, and home care services specifically, comprises a significant portion of the growing service sector in the US and elsewhere, we must understand how workers manage the constraints of their employment and to what extent they craft a sense of ownership, satisfaction and dignity on the job.

Labouring in home care

Home health is a growth industry, especially in the areas of personal care and homemaker services (Bureau of Labor Statistics 2004, Stone and Weiner 2001). Unlike institutional care (*i.e.* nursing home or convalescent care), home care in the US is an amorphous industry composed of a complex array of publicly-funded, not-for-profit social services and for-profit agencies (Benjamin 1993, Stone and Weiner 2001). In California and in most of the US, both skilled and unskilled providers work in home care. Public health nurses, home health nurses and registered nurses work as skilled providers, usually tending to patients for a limited period of time after a stay in hospital. These workers tend to work for a hospital, Health Maintenance Organization (HMO) or agency, providing hands-on medical care to clients, such as post-operative care, changing catheters or administering IV drugs. Pay and benefits are relatively high for this professional group. Home health nurses, for example, receive a median annual income of $45,890 (Bureau of Labor Statistics 2004).

Unskilled providers working in home care include personal home care aides and home health aides. Home *care* aides – the subject of this study – generally work on their own, with only periodic supervision from a nurse or social worker. As a result, aides are strictly limited to non-medical care of clients. Generally, they provide housekeeping and personal care services, such as bathing, dressing, meal preparation and companionship. Home *health* aides, in contrast, tend to work directly with a registered nurse as she visits clients in their homes. Neither position – home care aide or home health aide – requires formal general education or medical training, although some agencies require CPR certification, evidence of basic first aid skills or other minimum qualifications. Aides may also opt (and are sometimes required) to train as a Certified Nursing Assistant (CNA), which amounts to about 150 hours of training and competency evaluation (Ong *et al.* 2002).

The pay of home care workers varies, based on region, the employer and whether the employees are unionised. The median hourly wage of a home care aide in the US is $7.81/hr, with the lowest 10 per cent earning less than

$5.90 an hour and the highest 10 per cent earning more than $10.67 an hour (Bureau of Labor Statistics 2004). Recent labour victories in California, New York and Illinois represent an important shift in the bargaining power of home care workers in the US. For example, in California, wages for state home care workers rose to an average of $9.00 per hour in 2002 after union-isation, from a wage that hovered around $7.00 an hour only two years earlier. While these changes reflect a significant increase in pay, entry-level earnings still fall below the federal poverty level and in many places, like California, the pay fails to constitute a living wage (Howes 2002, Ong *et al.* 2002). Benefits such as health insurance or retirement pay are rare, especially for part-time workers (Ong *et al.* 2002).

The lack of standardised pay, health benefits and overtime compensation for home care workers in the US stems from confusion and ambivalence – on the part of the state, workers and employers alike – about whether the act of caring for someone in their home qualifies as waged labour. For example, the *Fair Labor Standards Act* (*FLSA*), initially passed by Congress in 1938 and amended in 1974 to include domestic workers, continues to exempt certain types of domestic caregivers from the protections of the Act, namely minimum wage and overtime pay. Although nurses and even house-keepers are protected by the Act, most home care aides are exempt because they are seen as 'companions' to their elderly or disabled clients (Biklen 2003). Although workers have begun to challenge this narrow interpretation of the *FLSA*, they nonetheless continue to provide care under highly variable conditions, depending on where they live, how they are employed and whether unions have succeeded in organising workers in the area.

Method and description of sample

This research is based on a grounded theory approach (Glaser and Strauss 1967, Strauss and Corbin 1990) to qualitative data collection and analysis. Data are drawn from a year-long qualitative study of home care services in Central City, California, a medium-sized city with approximately half-a-million residents. In-depth interviews were conducted with 33 home care workers: 23 worked for a state-funded social services agency, California Home Services (CHS) and 10 were employed by a private agency, It's For You! home care. Particular emphasis was placed on CHS workers because the programme is the country's largest state-funded home care employer, with just over 200,000 workers (Schneider 2003). Central City, California Home Services (CHS), and It's For You! are pseudonyms used to protect the confidentiality of study partici-pants. Written and oral consent was obtained from all respondents.

To ensure validity of results, the interview data were triangulated with on-site observations of home care (approximately 30 observations), union meetings (2), worker trainings (20), new employee orientations (2) and worker appreciation ceremonies (2). In addition to interviews with 33 home care workers, I interviewed nine public health nurses, three county social workers and five agency administrators/proprietors, all of whom worked in home care in Central City. Permission was granted by both the state agency and the private home care company to do the research, although I was restricted from formally recruiting and interviewing older people or disabled clients (informal questioning was permitted).

Observations of home care 'on-site' took place in two ways: during site visits with public health nurses, and in conjunction with interviews of home care aides. For six months I shadowed public health nurses as they visited elderly or disabled CHS clients with escalating health problems. During each site visit, I introduced myself to clients and home care workers as a sociologist learning about home care. Nurses generally conducted two or three home visits a day, each one lasting between one and two hours. 'Jotted notes' (Lofland and Lofland 1995) were taken on site and then expanded into full fieldnotes at the end of each day of observation. On a few occasions, on-site observations with the nurses led to a follow-up interview with a home care aide.

More often than not interviews with home care aides were secured at caregiver trainings, orientations or union meetings. After gaining permission from the event organiser, I announced my research to the room and asked for study participants, sending around a sign-up sheet for those interested. These events yielded the 23 interviews conducted with CHS (state employed) workers. Although as many as 25 caregivers would sign-up for interviews during a given meeting, most declined to be interviewed on follow-up, even though I offered a $20 stipend to each study participant. It's For You! (private agency) workers were recruited over the phone from a complete list of home care workers obtained from the agency and from flyers posted in the agency office. Of the approximately 30 active aides on payroll, 10 agreed to an interview.

Home care workers were interviewed in client homes, their own home or a public place. When I interviewed home care workers in client homes, an elderly or disabled client was usually present, although in most cases the client was asleep in another room or did not appear sufficiently mentally competent (or interested enough) to follow the conversation. Written and oral consent was obtained from workers before the interview. More often than not, caregivers generously opened up their homes (or places of work) to me; many spent time giving me a tour of their home or 'workplace'. Interviews with home care aides lasted anywhere from one and a half hours to three hours, depending on the location. Interviews with caregivers 'on the job' lasted anywhere from three to five hours, as part of the time was spent observing caregivers as they tended clients or as they completed household tasks.

After transcribing all interviews verbatim, data were analysed using EthnoNote, a Qualitative Data Analysis (QDA) software tool that helps to organise, search and sort data by code. Although the interview sample yielded a wide variety of caregivers, with a range of motivations and levels of commitment, the fact that I sampled caregivers from voluntary trainings and orientation meetings means that workers who take their care responsibilities seriously are over-represented. That said, workshops attract caregivers through incentives of free dinner, gifts or money, which suggests the pool of caregivers might represent a somewhat broader range of caring commitments and motivations among this population of workers.

Home health aides discussed in the chapter varied in the number of hours they worked, their work experience and training, and the extent to which they were the primary (as opposed to secondary or tertiary) caregiver for a client. The hourly wages of workers interviewed ranged between $7.25 per hour and $9.25 per hour. The participants all lived and worked in Central City, California and ranged in age from 21 to 70 years. The median age of the home care worker was 52 years. The racial composition of the sample of home care workers was as follows: 36 per cent (12) White, 27 per cent (9) Black, 18 per cent (6) Latino, 12 per cent (4) Asian and 6 per cent (2) Other. Of home care aides interviewed, 82 per cent (27) were female and 18 per cent (6) were male. The sample roughly mirrors the general population of home care workers in California and the US, with a high percentage of women and an over-representation of racial and ethnic minorities (US Bureau of Labor Statistics 2003). The breakdown of clients by race/ethnicity matches the breakdown of home care aides: of the 33 aides interviewed, only four were caring at the time for a person of a race or ethnicity different from their own.

Literature review: dignity at work

Recent scholarship in the sociology of work and occupations focuses on the innumerable constraints facing low-wage workers in the new capitalist economy (Ehrenreich 2001, Hodson 2001, Munger 2002, Sennett 1998, Smith 2001). Consisting largely of poorly-compensated service sector jobs, the current work environment demands increased flexibility on the part of workers who must contend with decentralised workplaces, part-time or contingent work arrangements and worker-participation programmes promoting 'team building' and increased worker initiative. While the demands for employee flexibility have increased, the traditional material rewards of work – a living wage, benefits and the right to union representation – continue to be eroded.

Critics of the new capitalist economy argue that lack of financial security is not the only constraint facing workers today. Richard Sennett (1998) makes the case in *Corrosion of Character: The Personal Consequences of Work in the New Capitalism* that it is nearly impossible for workers to form a coherent identity on the job, due to the impermanent, insecure nature of contemporary work arrangements. Sennett conveys the important point that the trend toward so-called flexible employment compromises not only the financial wellbeing of workers, but also affects their selfhood. In essence, the worker pays the price in character, while the employer benefits in profit.

While Sennett provides a timely and critical portrait of the changing nature of work, recent scholarship in sociology extends his account to consider how workers themselves experience and understand new employment arrangements. Sociologists Randy Hodson (1991, 2001) and Vicki Smith (2001) emphasise the importance of empirically-grounded accounts when assessing the implications of constraining work environments. Greater attention is given to worker agency in a grounded approach, so as to assess – rather than assume – the deleterious effects of the new economy on workers. Hodson (1991) in particular takes issue with the 'theoretical straightjackets' that tend to limit critical studies of work. In his estimation, radical perspectives tend to explain worker behaviour either in terms of 'acquiescence (false consciousness) or resistance to capitalist control of the workplace' (Hodson 1991: 48). Unsatisfied with either view, Hodson proposes a study of work that pays close attention to the social contexts and social interactions that inform the work experience. Recalling the classic work of Everett Hughes (1958, 1971), Hodson seeks to understand how workers – through interactions on the job with co-workers as well as clients or customers – attach meaning to their labour, often finding autonomy and dignity in a job that outsiders perceive to be thankless dirty work (Hodson 1991).

Recent ethnographies of work in the service sector take up Hodson's call for grounded accounts of worker experience. Katherine Newman (1999) argues that low-waged workers in the inner city derive a sense of dignity from their work as fast food employees, primarily because they take pride in making a conscious choice to avoid unemployment, street crime or dependence on welfare. Although the jobs are poorly compensated and demanding, Newman suggests that workers borrow from middle class sensibilities of responsibility to import meaning into the work and to extract as much as they can from the job, whether in terms of job skills or a sense of fulfillment. Smith (2001) looks at a broader class of service workers in the US and finds that workers in the three companies she studied were willing to accept economic uncertainty because 'they felt they were gaining skills and insights that would allow them to maintain a solid footing in the new economy' (2001: 9). In other words, Smith finds that even though the contemporary workplace is indeed uncertain, workers find ways to manoeuvre within these constraints.

Low-waged healthcare work

Low-level healthcare jobs comprise a significant portion of the growing service sector and yet relatively little research in sociology considers the organisation of these jobs and their impact on workers. Existing health service research on low-skill health workers is generally quantitative and emphasises the macro-structural conditions that cause provider turnover and burnout (Crown 1994, Hollander Feldman 1994, Stone 2001, Yamada 2002).

The few ethnographic studies of unskilled health providers tend to focus on workers in institutional settings such as hospitals or nursing homes. Tim Diamond (1992) and Nancy Foner (1994) both examine how nursing home aides manage a job that takes place in an increasingly bureaucratic and profit-driven environment. Of particular interest to the authors are the ways that aides struggle to maintain a sense of ownership and autonomy in a context where corporate and middle management restrict the amount and quality of care given to older clients. Diamond (1992) finds that while nursing home aides become distressed and jaded by the work conditions, they collaborate with clients to find ways to subvert management control, thereby maintaining a sense of autonomy and dignity over their work. Foner (1994) finds workers respond similarly to the 'speed-up' of care but suggests that while autonomy may give aides a sense of ownership and dignity on the job, it is possible that such autonomy may be misapplied to the detriment of client wellbeing. Both Diamond and Foner provide rich ethnographic examples of how low-skilled health workers manage to create dignity and autonomy on the job, in spite of bureaucratic controls on their labour.

It is curious, given this increased interest in the questions of bureaucratic constraint and worker autonomy among nursing home aides, that the analysis has not been extended to home care workers. Deborah Stone (2000), in one of the few qualitative studies on unskilled home care workers, concludes that care-givers have a great deal of freedom to direct care, but rarely misapply this auto-nomy to the detriment of client wellbeing, as Nancy Foner portends. Stone finds that home health workers go out of their way to spend extra time with clients and pay out of pocket for client expenses, thereby preserving their own ethic of care. Aronson and Neysmith (1996) take a more critical view, suggesting that the interpersonal bonds that develop between caregivers and their clients obscure the fundamentally exploitative nature of the arrangement.

While each of these authors makes an important contribution to the nascent literature on home care work, the tendency is either to romanticise the importance of affective ties between worker and client, or to overstate the exploitative nature of the relationship. I argue that the reality of the work experience of the home care provider lies somewhere between the two positions. On the one hand, workers recognise and lament the constraints of their low-wage labour; on the other, they find reward in a job that allows them to experience some autonomy and creativity through their carework. I suggest the rewards that come from caregiving allow workers to import dignity into a highly stigmatised ('dirty') occupation. Dignity is most salient for those workers who discover home care after fleeing a particularly alienating service job, either within or outside the healthcare industry. For these workers especially, home care becomes an important point of connec-tion to their labour, rather than a source of alienation.

Findings

Grounded accounts from home care workers in Central City tell a complex and contradictory story about the labour of caregiving. Aides are quick to identify the constraints that make their work difficult, including overwork and added responsibilities, increased risk on the job, and the physical and emotional strain of providing care. At the same time, workers talk at length about the dignity they derive from working with clients. Based on the interviews and observations, I identify three sources of dignity in home care: practical autonomy on the job, especially relative to previous work in the service sector; skills building; and pride and honour in dirty work.

'We're Maids Plus': overwork and added responsibility
Like many service jobs in the US, the work of the home health aide requires a great deal of emotional commitment and flexibility, at the same time that it offers very little in the way of job security or wages. In my sample, 23 of 33 aides discussed the stress that came from having multiple jobs for multiple clients to ensure enough weekly income simply to pay their own bills and rent. All caregivers agreed that they should be paid more for what they did. In general, when I asked aides why they thought they were underpaid, most cited the undervalued nature of the work. In general, caregivers minimise the low-pay and insecurity of the work as 'part of the job'.

Katy, a mixed-race white and Filipina woman who has worked in home care for over 30 years, now accepts that 'outsiders', including doctors, nurses and family members, fail to recognise the contribution of home care aides. Over lunch at a coffee shop, we discussed the lack of respect she has faced over the years:

> Nobody'll listen to you . . . you're just the aides. I get so tired of being thought of as incompetent and stupid and don't know anything. But I think it's always been that way. And I think it will continue to be that way. We're the ones that know the patients, and everything. But it's the power trip, the control trip . . . I could care less. I do what I do, you know?

Aside from the obvious drawbacks of working in a low-paid, undervalued job, home care aides identify overwork and added responsibilities as significant constraints of the job. Aides who care for a client and feel a sense of loyalty to him or her, often find themselves in a situation where they are being asked to stay a little longer, lend a little money or take on a little more cooking and cleaning. Rosa, a recent immigrant to the US from Mexico, supports two sons on her income as a home care aide, and while she loves the work of caring for people, she often contemplates leaving home care to work as a house cleaner because the pay is better and the hours more finite. As Rosa put it when I asked her to describe the drawbacks of her carework: 'We're Maids Plus, you know? Maids plus companion, maids plus nurse, maids plus family'.

Sophie, a Hungarian woman in her fifties who recently immigrated to the US, spoke of the expectations that were placed on her in a recent home care job where she effectively became 'one of the family'. She explained how the blurring between work and family could create a situation where overwork simply became part of playing the role of fictive family member:

> Now in this family, they made me kind of part of the family, and the lady was very nice, but with being part of the family they expected me to be with her when the daughter was out of town on Thanksgiving. I said, 'well I would like to visit my husband's family in Southern California'. The lady said, 'well I will need you because my daughter's out of town and I need you'.

As is the case for the domestics discussed in Hondagneu-Sotelo's (2001) *Domestica*, the rhetoric of the 'family bond' in home care often masks the inequality and exploitation of the carework arrangement. Certainly, the overwork demanded of aides stems, in part, from the fact that carework generally blurs the line between informal and formal labour (Folbre 2001, Glazer 1993, Harrington Meyer 2000). Kelly, a white sixty-year-old aide, often let her client sleep at her home, even though the terms of their contract forbad this. Kelly was paid for three hours of care a day, but often spent a week of uninterrupted time with her elderly client. She was aware of breaking the rules, but said that both she and the client liked the companionship and that the client was afraid to stay alone at night in her home. Kelly explained that 'It's not an issue, we're friends. We enjoy each other's company and we go out to eat together . . . I don't need a whole lot of money'.

Kelly's willingness to take on extra, uncompensated work to help her elderly client makes sense when considering her recent history as an unpaid caregiver of a son who had died of AIDS and a husband who had died two years later after battling cancer. Although Kelly acknowledged that she struggled to make ends meet and that she should be paid more for the work, she also seemed to accept the situation, as it closely resembled her prior work as an informal, unpaid caregiver.

For other home care aides, caring for someone means taking on the added responsibility of paying for living or medical expenses. Given that many clients, especially in the CHS programme, are often as poor as the caregivers, it is not surprising that workers become involved in helping to support a client who, in many instances, becomes a friend. Martina, a 60-year-old African-American aide caring for a disabled African-American man in his fifties, recounted a story about paying for her client's medication because it was no longer covered by insurance:

> We went to the neurologist [who said], 'well, I'm going to give you some lidocaine pads'. I went to the pharmacist and the pharmacist said, 'they're not going to pay for this'. They were $175 dollars, so what are you going to do? Then after I bought them and we put them on his foot . . . he said it froze it and he didn't like it. So now I'm stuck

with $175 dollars worth of lidocaine pads. And the Vicodin he takes, Medi-Cal doesn't pay for them cause they're too strong . . . those I paid $78 dollars for.

Andrew, a 55 year-old African-American caregiver, also absorbed some of the costs associated with his client's care. The client was a young, schizophrenic African-American man 'living-in' with Andrew and his wife because he required full supervision. Andrew's contract specified that the client must pay him over $500 for rent, although when I interviewed Andrew, he had recently reduced the rent commenting, 'I was seeing that he was struggling with that, he was having real problems'. Although Andrew's generosity is to be commended, the larger question is why someone like Andrew, as a low-income caregiver, absorbs financial responsibility for a client whose federal disability allowance fails even to cover his rent. Once Andrew's client had settled into his home, it became difficult for Andrew to remain rigid about the terms of the work contract, an example of how a blurred boundary between formal and informal labour helps to sustain overwork and added responsibilities among caregivers.

Risks of providing care

While agencies are careful to remind aides during trainings that they are forbidden to provide medical services like changing catheters, dressing bandages or injecting insulin, in practice agencies seem to take a 'don't ask, don't tell' approach to monitoring aides. The aides I spoke with and observed on the job were aware that the inattentiveness of agencies shifted responsibility – and risk – into their untrained hands. In response to this risk, some aides maintained rigid boundaries and refused to take on medical tasks, while other aides accepted – even welcomed – the risk as part of the job.

Jennifer, a white caregiver who worked for an elderly woman on the outskirts of Central City, was given a great deal of latitude to monitor and treat her client's diabetes, even though she was not completely comfortable doing so. She recounted a story of the time her client's nurse 'ordered' her to 'use her discretion' when supplying the client with insulin, even though Jennifer expressed concern to the nurse that such decisions were beyond her expertise. Eventually, Jennifer realised no one else was going to provide the care, so she cautiously accepted the responsibility, learning over time to manage her client's disease.

Jennifer displays some hesitation to take on work she is not trained to do, and her caution is not unwarranted: if something were to happen to her client, Jennifer could be held liable. CHS, the state agency that pays Jennifer's wage, does not assume responsibility for caregiver behaviour, arguing that the contractual relationship is between the client and the caregiver, not the caregiver and the agency. Private agencies like It's For You! do contract directly with the client (and then 'assign' a caregiver to the client) and, as a result, appear more diligent about reminding aides that they should not provide medical care for clients. Nevertheless, home care aides from both agencies in the study report having to occasionally provide care outside their skill set.

Andrew explained that helping clients with their medical needs came naturally to him, after years of watching and learning on the job. He suggested that the agencies were fully aware of the fact that aides often did work they were not formally trained to do:

> It's not in agreement with the agency but they know we do a lot of things personally for our clients, cause once you're with a client for four or five years, you establish a kind of rapport with them, a friendship and trust. I had one [client], BD, he's a very wealthy man, and I did just about everything for him.

Mark, an Asian-American caregiver in his forties, claimed that in the CHS programme aides often provided more hands-on care than public health nurses, even though PHNs were some of the most highly-trained professionals in the field:

> Aides can get away with more things because they do not have the censorship or the people that they have to report to all the time. Literally and legally, if you had four and five different medications that you had to take, I could literally lay 'em out and have you take 'em. And maybe even illegally but legally place them in your mouth because you needed help. Whereas an RN is not qualified to do it. Even though I'm not [an RN], I could get away with it and I'd have more range and variance, whereas an RN, they have to be put under constant microscope, having to report to everything and everybody. And so basically we can do almost the same things they can to a certain degree, we just can get away with it without the training.

Although Mark seems confused about the work responsibilities of nurses in home care – they can in fact administer medication orally – he nonetheless touches on the strict guidelines limiting nurses in the field. During my observations, CHS public health nurses expressed frustration at being unable to provide direct medical care to patients. Shadowing one nurse, I watched as she instructed an elderly, nervous diabetic client how to take an insulin reading. Unable to touch the client – CHS nurses cannot touch patients in situations where there is blood or an open wound – the nurse could only stand and watch as the client fumbled over the task. Limited in practice because they are not directly under a doctor's order, CHS public health nurses acknowledge that aides often absorb responsibility for minor medical care of clients, simply because there is no one else to do it. Although nurses resent their shrinking jurisdictional boundaries in home care, it is the clients and home care workers who truly absorb the risk of this arrangement. Aides run the risk of losing their jobs if they 'get caught' or if they harm a client;

and clients are often forced to choose between substandard care from an aide or none at all.

Given the risk to both aide and client, it is not surprising that some home care workers I interviewed refused to take on medical care of clients. Fahima, a Belizian-born caregiver who had lived in the US for 10 years, refused to jeopardise her job by performing tasks outside her expertise. She explained:

> If they have a wound that has a band-aid or something like that, I'm going to take that off and clean it with peroxide and put it back on. But if they have a big open wound, I'm not going to do that. And I let them know, I'm not going to do that, I'm not going to change no IV for you. I'm going to stick to what I'm supposed to do. And that's it. If you don't want it, that's just too bad. But I'm not going to cross the line. I'm not going to jeopardise my job to please nobody.

While Fahima's clear boundaries protect her (and arguably the client) in the long run, many aides describe feeling 'torn', because if they do not provide the medical care, the client often goes without. Given that many aides feel 'like family' to their clients, it makes sense that some are willing to absorb risk associated with this type of care.

Physical and emotional strain

The risks of working in home care go beyond the fear of harming another or of losing one's job. Far more common are the risks associated with the physical and emotional strain of providing direct care to another person. Home care and nursing home aides, for example, suffer from the highest number of musculoskeletal disorders of any occupational group in the US (Bureau of Labor Statistics 2004a). Given that many clients are unable to move without the help of an aide, transferring a patient from one place to another becomes an important – and dangerous – requirement of the job. After I attended a 'transfer' course with a group of caregivers, it became clear that even the smallest aide can safely lift the largest client if properly trained. However, the training courses are optional and many aides cite lack of time as the reason why they had not learnt the proper technique for transfers. As a result, approximately one-third of the sample (10 aides) reported some kind of job-related injury, although only one had notified her employer.

Joyce, an African-American caregiver in her sixties, had gone into home care after losing her life-long job delivering phone books for a local telephone company. At the time of the interview, Joyce cared for her brother with cancer, but explained how she contracted a back injury while caring for an dependent woman over two years previously:

> This particular morning I'd sat her on the potty chair, but then instead of her going to the potty chair she went the other way, which means that I

had to go the other way and pull some muscles in my body. As an end result of that I didn't work any more. I went to a doctor and of course the insurance bucked, I mean really bucked. The doctors did not act on my behalf because they wrote a note saying that I could go back to work. I could not go back to work, and I did not. It cost some 300 dollars for me to go there [chiropractor] for an hour, and I'm saying, wait a minute, he's not doing that good. So I stopped going to physical therapy and I still have problems with my back.

Sophie, a Hungarian caregiver, reported similar troubles transferring a large young man, disabled after a car accident. For her, the problem of moving a client was not insignificant; she left the job after realising that she did not have the skills to do the work safely. She described how her client leaned 'like a tree' and that she would always fall in the direction that he did. Eventually, Sophie quit because she realised that transferring her client was dangerous. Considering the severity of the client's disability and Sophie's reluctance to learn the proper transfer techniques, it is probably best for all involved that she did not continue to work with this particular family. However, it is not difficult to imagine a scenario where a caregiver might persevere through such physical difficulties because of the need for work, even though she might potentially harm herself or the client.

In addition to physical strain, home care aides also identify the emotional toll produced by this type of carework. Emotional strain comes from failing to connect with a client, as well as from emotionally over-investing in a client's wellbeing. Andrew spoke about the way his work sometimes made him feel disconnected, or 'robotic':

Sometimes he [the client] comes in with a big demand, you know, and he forgets that I'm human, he thinks I'm a robot. Like I say, it has to take a special type of person to do this kind of work because sometimes people, clients, they forget that you're human. They don't care. You have to remind them, 'Hey look, I'm a human being'.

Andrew's complaint that his client fails to recognise him as human reflects an alienation of self most commonly associated with emotion work. As Hochschild (1983) reminds us, however, emotional labour can also be a problem when a worker over-invests too much of the self into work, such that burnout from emotional involvement becomes likely (Hochschild 1983, Wharton 1999). While the cost to workers in this scenario is not as high as compared to situations that produce an alienation of the self, the emotional toll of over-involvement is still significant (Hochschild 1983). Confirming Hochschild's assertions, home care workers in this study spoke of the emotional costs associated with investing so much in the care and wellbeing of another person. Luis, a Puerto Rican caregiver, explained how he felt when a client passes away:

You get close to people. First thing they taught me when I was in school was, don't get attached. But, it's a lot of malarkey; it doesn't work that way, not in real life. You could tell somebody that but when you're working, you get close to somebody, and when that person passes on you feel it. They say that it gets easier to accept, but every once in a while you're going to have someone who's special to you and it's going to hurt.

Lupe, a Mexican American home health aide, had a very conflicted emotional relationship with her client. The woman for whom she cared, a white woman in her nineties, was very demanding, deriding Lupe for 'eating too much junk food' or for 'being fat' at the same time that she begged Lupe to stay with her. Lupe seemed torn between feeling a strong connection with her client and fearing her rejection. In many ways, the dynamics that emerge are reminiscent of those between unpaid family caregivers and their kin (Karner 1998), producing many of the same effects of burnout and emotional over-involvement.

Andrew also talked about the tumultuous caregiving relationship, describing how he would 'clash and argue' with clients but ultimately become reconciled. In one case, Andrew did leave his client, saying that it was 'like breaking up from your wife or something'. Andrew admitted to feeling over-involved with other clients, likening the bond to 'being married . . . not like being gay, but it's like your family'. It was not uncommon for caregivers, like Andrew, to emphasise both the benefits and drawbacks of the over-involvement that comes from close ties with another person. In some cases, the emotional toll is so great that caregivers ask the agency to place them with a different client. One can also imagine how such emotional over-involvement might result in burnout and an inclination to leave the occupation altogether, a supposition supported by recent evidence suggesting significant turnover in the home care workforce (Ong *et al.* 2002).

Finding dignity in dirty work

Although home care workers identify several ways that their work is constraining, they also talk about the rewards of caregiving. I identify three sources of reward for aides: practical autonomy on the job, especially relative to prior work in the service sector; skills building; and doing dirty work. The first two rewards – practical autonomy and skills building – can be interpreted as individual rewards, or rewards that allow workers to achieve a sense of ownership and control over their labour, two factors central to maintaining dignity on the job (Hodson 2001, Wardell 1992). For home care aides, however, there is also a third reward – doing dirty work – that is unique to the caregiving relationship, and is relational (rather than individual) in nature. Workers draw meaning from their willingness and ability to perform dirty and mundane tasks that others avoid, knowing that their efforts improve the lives of clients. Taking on dirty work, therefore, is an important source of dignity for workers whose labour is invisible and undervalued by the general public.

Practical autonomy

Agencies generally dictate – at least on paper – how many hours home care aides may cook and clean, and how much time should be spent on bathing, dressing and running errands. However, because workers are generally not supervised on a day-to-day basis, they report taking as much or as little time as needed for a task without much regard for the bureaucratic dictates of the job. The fact that caregivers are free, relative to their colleagues in nursing home care, to use their own discretion when assigning time to tasks, translates to a form of 'practical autonomy' on the job (Wardell 1992). Distinct from professional or absolute autonomy, practical autonomy is a way that workers create and manage their own environments within certain constraints. Wardell (1992) argues that most employers overlook, or even encourage such autonomy, precisely because it allows the business enterprise to proceed smoothly. Home care workers welcome practical autonomy because it gives them greater control over their labour, a significant factor in attaining dignity in the workplace (Hodson 2001).

Home care workers interviewed describe a sense of practical autonomy that comes from having the freedom to informally negotiate certain terms of employment with clients, depending on client needs and level of functioning (*i.e.* what time to arrive and depart; when to perform certain caregiving tasks). In both the context of CHS and for-profit home care, all aides – in some way or another – seek out and hold on to jobs where they have some flexibility to determine hours worked and control over care. Some workers describe this as a sense of 'being your own boss', while others simply like the fact that they can work with the client to determine when and how to provide care. Of course, practical autonomy is also linked to the earlier discussion of risk. While workers can exercise some creativity on the job, this 'freedom' often means they must assume the risk of taking charge of medical care outside their formal training. Aides seem conflicted about this tension: like most of us, they seek jobs that allow for creativity and autonomy because of the sense of dignity they provide; but with this creativity and autonomy often come unwanted – and risky – responsibilities.

The most common way home care aides describe the value of practical autonomy is to compare their home care work with previous experiences in the service sector. Of 33 aides, 32 had worked in other kinds of service jobs, including flipping hamburgers at McDonalds, working as janitors or domestics or providing care in institutional settings. Several workers sought out home care as a way to flee unsatisfying or inflexible work environments, while others 'fell into' carework more informally (caring for a relative or friend) and then realised that they could get paid for this type of work. One constant among respondents who left other service jobs and moved into home care was the report of increased autonomy and decreased stress. Although respondents admitted carework was very stressful at times, most made clear that other service work was even more so.

The rewards of autonomy are most salient for those aides with previous experience working in an institutional care setting. Without exception, the 19 (57%) aides in the sample who had had prior experience in nursing homes or mental health facilities spoke about the poor working conditions and the depressing nature of the work, likening facilities to 'factory assembly lines'. Most recounted emotional stories of working eight- or ten-hour shifts without a break and feeling as though no matter how hard they tried, clients did not receive the proper care. Much of what the workers lament is the lack of autonomy to provide the kind of care nursing home clients need, a sentiment found in Diamond (1992) and Foner's (1994) ethnographic accounts of working conditions in nursing facilities.

Luis, a Puerto Rican man working in home care, claimed he would never go back to institutional care, where he had worked for 20 years, because of the poor working conditions and low standards of care:

> A lot of places, the workers don't have the time, there's not enough time in the day to do it. Me, I didn't have much time either because I hardly got to lunch. I had maybe a 15-minute break in my whole shift because I had to make sure these people were right. Taking into account that they are going to give low man on the totem pole all the patients that no one else wants. It's hard for you to dedicate enough time for each one of the residents.

Camilla, a 42-year-old African American aide, expressed a similar sentiment about her experience working in a nursing home, emphasising that the client also suffered when care was 'sped up':

> To me it's too busy. It's not enough time for the client. You know what I mean? You have no personal time with them. You are going to give them a bath real quick, check their temperature, blood pressure, and you're out of there. What about 'How do you feel today?' or 'Did you sleep well?' 'Did you have any dreams?' 'Is there anything bothering you?' You know, rub their head and take time to do all that. The important stuff, that's what I do.

Caregivers like Camilla and Luis welcome the practical autonomy of home care precisely because they are relatively free to care according to their own standards, without working under bureaucratic management that encourages 'speed up' of services (Diamond 1992). Considering their prior experiences in the service sector, especially the nursing home industry, practical autonomy becomes a key mechanism through which home care workers maintain greater control over their labour (and therefore the care of the patient). Although there are potential drawbacks to increased worker autonomy for both caregiver and patient (Foner 1994), autonomy also promotes a sense of ownership over the standard of care and, by extension, a sense of dignity for the worker.

Building skills in home care

A common view of home care – among doctors, nurses, family members and the general population – is that it constitutes a 'dead-end' job that anyone can do. Indeed, a public health nurse commented that agencies accept 'anyone off the street', which leads, in her estimation, to a workforce that cares little about the skill or craft of caregiving. While there is no doubt that a certain percentage of home care aides simply log their hours and go home, aides I interviewed view home care as an occupation that provides them with a unique caregiving skill set. A subset of these workers also emphasise that they intend to put these skills to use in other careers, in and outside the healthcare field. Irrespective of whether skills building provides home care aides with horizontal or vertical mobility within the job market, the perception of skill is central to the way that aides craft dignity on the job, countering misperceptions that caregiving is work 'anyone can do'.

Jackie, an African-American caregiver, believed she had perfected the skill of bathing clients, something she now 'specialised' in. When I asked her to elaborate, she responded:

> Let me make this note. I'm going to tell you, if you ever meet one of my clients, they would tell you, I give them the best bath, shower, even the men. Everybody does not know how to give a bath. I wish I could teach 'em how. I don't want to sound racist or nothing. The white women that I've worked with and I've seen give a shower, they don't know how to do it. They do not know how to give a shower.

Jackie considered herself something of an expert when it came to bathing and showering, a skill she indirectly attributed to her racial and cultural background. Jackie is not alone in her assertion of a 'cultural skill' or predisposition to care. George and Hannah, a married Filipino couple who immigrated to the US on a work visa sponsored by the agency, feel that Filipino workers are more 'industrious' than American workers and that they have 'that thing' that makes them attractive caregivers. The couple also feel they have skills as caregivers that doctors and nurses do not have, claiming that 'nurses only give medications and do none of this intimate or personal thing with the patient'. Such 'boundary work' (Lamont 1992) is not uncommon among caregivers, who draw a sense of dignity from the belief that they possess skills that more accredited providers do not possess.

Jennifer, a white aide caring for a frail elderly woman, viewed caregiving skills as part of her 'nature', but she also saw her work in more pragmatic terms. She described her job as a 'stepping stone' that would lead to other opportunities in home care, such as owning her own small residential home for the elderly. Jennifer attended training workshops regularly, actively building her skill set in the hope of creating opportunities for herself down the road. Other caregivers saw their 'soft' skills, such as inter-personal communication, as key ingredients to future success on the job market. Andrew discussed the importance of his ability to 'get along with people', something he learned from caring for clients of different races and backgrounds:

> I learned a lot from dealing with other cultures and different backgrounds as well. I learned a tremendous amount about what people think about other people, and their background and race. You learn so much. Very educational. It's been quite an experience for me and it's opened up other avenues for me. I probably wouldn't even want to go into real estate now if I hadn't had this experience. I mean, I'm venturing out and communicating with people from all different races and walks of life.

Andrew described a new set of interpersonal skills that had come from his work in home care. In a sense, the interactive work of caring for people from different backgrounds helps build Andrew's human and cultural capital, skills that he hopes will translate into future job prospects. The perception on the part of workers that they are picking up important skills while caregiving helps in part explain why aides accommodate work conditions that, on the surface, appear constraining and against their best interest (Smith 1996). Whether these workers go on to achieve long-term job mobility is an unanswered empirical question and beyond the scope of this chapter. It is clear, however, that in the short-term, aides view caregiving as a way to build job skills, thereby crafting a sense of dignity within the confines of a job often seen by outsiders as unskilled and menial.

Doing dirty work

Perhaps the most compelling – and counterintuitive – way that aides derive dignity on the job is through a sense of pride that comes from doing 'dirty work' (Hughes 1971). Home care is both dirty in the literal sense (taking charge of someone's personal hygiene) as well as in a broader Hughesian sense, referring to the mundane and repetitive nature of the work (Hughes 1971). As mentioned earlier, aides recognise their work as dirty, but don't necessarily internalise this stigma in obvious or predictable ways. As Everett Hughes (1971) reminds us, 'people do develop collective pretensions – to give their work, and consequently themselves, value in the eyes of each other and of outsiders' (1971: 340).

One way that aides import value and dignity into home care is to emphasise how well they have mastered the dirtiest aspects of the job. Katy, a caregiver nearing retirement, described herself as highly capable when it came to personal care (bowel and bladder care). Over lunch, she explained to me that she was very comfortable with such tasks, something that distinguished her from informal family caregivers:

You know, families don't cope well with it [personal care]. It gets into the couch and those kinds of things. Then I come over and clean it up . . . it doesn't seem odd to me. That's what older people do. It's about knowing that, for an old person, when you cough, you wet your pants. Wonderful. It's wonderful.

Few caregivers express as much enthusiasm about bowel and bladder care as Katy. However, nearly all take pride in the fact that they provide a valuable service that most people won't even talk about, let alone perform. Luis, for example, had worked as both a home care aide and nursing home aide. He took great pride in the cleanliness and appearance of his clients. It was clear that he viewed hygiene to be of utmost importance, because good hygiene humanised his clients while also marking him as an exemplary care provider:

My men, I don't like to see my men with any kind of, you know, shabby looking beards. Sometimes I had a gentleman that might want to grow a beard – that's okay – you can grow it, but we're going to keep it trimmed. The minute I see that it ain't working, it's coming off. We don't want you looking shabby. It just made a difference to all of them, you know, I mean their personal care. It's just hygiene, getting them dressed every day in something different. It could still be clean and I would say, 'No, I want to put something different on you today'. That's what they need because it's something that they used to be able to do but they can no longer do it. A lot of places, the workers don't have the time.

Providing daily care to clients – however dirty the work – translates into a sense of pride and even moral authority vis-à-vis other providers in the medical community. Home care workers are clearly at the bottom of the medical hierarchy and, not surprisingly, they report being disrespected or ignored by busy doctors or nurses. Even so, aides maintain a sense of dignity by distinguishing themselves from hurried, out-of-touch medical professionals. Camilla, an African-American caregiver, expressed her feelings in this way:

When you're a CNA [aide], that's the bottom they say. But you spend more time with the clients. And that's why I'll stay a CNA for a little while, because with the other ones – LVN and RNs – they do a lot of paperwork. They don't spend their life with people. They don't know. They come to the CNAs to ask about the patient, because they don't know. I'm like the number one person. I spend a majority of the time with them.

Camilla emphasised the unique bond that was formed between aides and clients, highlighting the relational nature of the work. As scholars of carework note, this 'relationality' is not a one-directional relationship of dependency.

Rather, both caregiver and care recipient draw value – and I would argue a sense of dignity – from constant, and often very intimate, social interaction (Kittay 1999, Parks 2003, Tronto 1993). Lupe, for example, worked nights as a packer for Target, in addition to her work as a caregiver. Lupe was always tired from working two jobs and I asked her why she kept the caregiving job, given that it had no benefits, paid a lot less than her packing job at Target, and caused her so much emotional distress. She acknowledged that the stress of it all took a toll but that she kept with it because, 'it [caregiving] makes you feel good, like nothing else'.

Similarly, Luis had worked in home and residential care for nearly 20 years, earning as little as $2.50 an hour early in his career. He explained why he stayed in home care even though the work could be repetitive and thankless:

Yeah, I mean, it just keeps you going. It gives you more to keep doing more . . . it just makes you feel like, oh, I'm doing something good. I'm doing something I want to keep doing.

For Luis, and other aides like him, pride and honour come from being able to affect positively the emotional state of clients, simply by providing for their daily needs. Luis' assertion that 'it gives you more to keep doing more' also suggests that he finds home care a place where he can invest as much emotional energy as he deems necessary and that this, in turn, provides him with a sense of dignity on the job. These findings support the recent work by Teresa Scheid (2004), who suggests careworkers would benefit from greater *support* for their emotional labour, rather than policies that regulate or restrict it. Without sufficient outlets for emotional labour, Scheid argues, the worker is prevented from making a meaningful investment of self in their work (Scheid 2004).

Home care providers appear to draw significant reward from doing emotionally intense carework, even though the work is dirty and mundane at times. Fuelling the dignity of dirty work is the sense, among aides, that they alone take responsibility for tasks that few others will do; work that directly impacts on the wellbeing of their clients. Combined with a new sense of autonomy – relative to previous work in the service sector – and the belief that they are building skills on the job, it becomes clear why some home care aides find dignity in the work of caring for an elderly or disabled adult.

Conclusion

As the population ages, in the US and elsewhere, the demand for home care services is expected to increase significantly (Bureau of Labor Statistics 2004a). Unskilled home care aides will be on the front lines of this burgeoning industry, confronted with the realities of poor pay, job insecurity and the low status of the work. While health services research makes clear that the conditions of labour must change to prevent burnout and turnover, little is known about how home care aides themselves perceive the constraints of their 'dirty work'.

I suggest here that workers have complex and sometimes contradictory relationships to their labour. Aides identify three broad work constraints that compromise their ability to do a good job or to experience their work as meaningful: overwork and added responsibilities; increased risk; and the physical and emotional strain of the job. While these represent significant constraints, workers interviewed also highlighted the rewards that came from experiencing practical autonomy on the job, building skills and doing dirty work. The dignity that workers draw from these rewards helps mediate the obvious constraints of a low-paying and demanding service sector job. In short, although many outsiders view the work as without skill or 'dead-end', workers interpret their responsibilities and contributions in a very different light.

Findings support recent claims within the sociology of work and occupations that workers create opportunity and satisfaction, even in jobs characterised by low pay and insecurity (Smith 2001, Hodson 2001). Dignity is central to the job satisfaction for the home care aide, a factor that is generally downplayed (or ignored) in overly deterministic accounts of work in the new economy. Of course, the implication of worker dignity is not entirely rosy, at least in the case of home caregivers. As I argue elsewhere (Stacey 2004) the dignity workers draw from caring labour obscures the structural inequalities faced by aides, making the task of unionising home care workers very difficult. Precisely because aides derive meaning from the bonds formed with clients, conventional approaches to organising workers – *i.e.* in relation to their material interests – is inadequate for a labour force that finds intense meaning in the non-material aspects of the work. In a country generally hostile to unionisation, it is imperative that future organising campaigns address the material needs of workers while also validating the caregiving relationship that serves as a foundation for their sense of dignity.

This research also raises questions about the meaning of emotional labour for healthcare workers. Conventional reasoning tells us that burnout among direct care workers is a result of the burdens of emotional labour. The home care workers in the sample, however, tell us they find relief in work that promotes, even fosters, deep emotional connections with clients. The finding is especially true for those workers who discover home care after fleeing employment in a nursing home. Data further confirm Teresa Scheid's (2004) assertion that burnout in healthcare organisations is a result of the suppression of a worker's emotional labour, rather than a product of intense emotional over-involvement.

Given these findings, one important empirical question remains unanswered: Does the dignity of the worker correlate with improved quality of care for the client? Future research must consider this relationship carefully, with attention to the ways that autonomy and dignity serve as important foundations for the caregiving bond.

Acknowledgements

Portions of this paper were presented at the American Sociological Association 97th Annual Meeting 2003 in Atlanta, Georgia.

I would like to thank the following people for their invaluable feedback during the writing and revision of this chapter: Zach Schiller, Magdi Vanya, Stuart Henderson, Jeff Sweat, Anna Muraco, Carole Joffe, Vicki Smith, participants of the IHPS Writing Seminar, and the editors and anonymous reviewers of *Sociology of Health and Illness*.

References

Aronson, J. and Neysmith, S. (1996) You're not just in there to do the work: depersonalizing policies and the exploitation of home care workers' labor, *Gender and Society*, 10, 59–77.

Benjamin, A.E. (1993) An historical perspective on home care policy, *The Milbank Quarterly*, 71, 129–66.

Biklen, M. (2003) Healthcare in the home: reexamining the companionship services exemption to the *Fair Labor Standards Act*, 35 *Columbia Human Rights Law Review*, 113.

Bureau of Labor Statistics (2004) *Occupational Outlook Handbook 2003–2004*. US Department of Labor.

Bureau of Labor Statistics (2003) Table 11: *Employed Persons By Detailed Occupation, Sex, Race, and Hispanic or Latino Ethnicity. Household Data Annual Averages*. US Department of Labor.

Crown, W. (1994) A national profile of homecare, nursing home and hospital aides, *Generations: Frontline Workers in Long-Term Care*, Fall: 29–33.

Crown, W., Ahlburg, D. and MacAdam, M. (1995) The demographic, employment characteristics of home care aides and other workers, *The Gerontologist*, 35, 162–70.

Diamond, T. (1992) *Making Gray Gold: Narratives of Nursing Home Care*. Chicago: University of Chicago Press.

Ehrenreich, B. (2001) *Nickel and Dimed: On (not) Getting by in America*. New York: Metropolitan.

Folbre, N. (2001) *The Invisible Heart: Economics and Family Values*. New York: The New Press.

Foner, N. (1994) *The Caregiving Dilemma: Work in an American Nursing Home*. Berkeley: University of California Press.

Glaser, B. and Straus, A. (1967) *The Discovery of Grounded Theory*. Chicago: Aldine.

Glazer, N.J. (1993) *Women's Paid and Unpaid Labor: The Work Transfer in Health Care and Retailing*. Philadelphia: Temple University Press.

Harrington Meyer, M. (2000) *Care Work: Gender, Labor and the Welfare State*. New York: Routledge.

Hochschild, A.R. (1983) *The Managed Heart: Commercialization of Human Feeling*. Berkeley: University of California Press.

Hodson, R. (1991) The active worker: compliance and autonomy at the workplace, *Journal of Contemporary Ethnography*, 20, 47–78.

Hodson, R. (2001) *Dignity at Work*. Cambridge: Cambridge University Press.

Hollander Feldman, P. (1994) Dead end work or motivating job? Prospects for front-line paraprofessional workers in LTC, *Generations*, 18, 5–10.

Hondagneu-Sotelo, P. (2001) *Domestica: Immigrant Workers Cleaning and Caring in the Shadows of Affluence*. Berkeley: University of California Press.

Howes, C. (2002) The impact of a large wage increase on the workforce stability of CHS home care workers in San Francisco County, Working Paper, Institute of Labor and Employment, UC Berkeley.

Hughes, E.C. (1971) *The Sociological Eye: Selected Papers*. Chicago: Aldine Atherton.

Hughes, E.C. (1958) *Men and Their Work*. Glencoe, IL: Free Press.

Karner, T. (1998) Professional caring: homecare workers as fictive kin, *Journal of Aging Studies*, 12, 69–73.

Kittay, E. (1999) *Love's Labor: Essays on Women, Equality and Dependency*. New York: Routledge.

Lamont, M. (1992) *Money, Morals and Manners: The Culture of the French and American Upper-Middle Class*. Chicago: University of Chicago Press.

Lofland, L. and Lofland, J. (1995) *Analyzing Social Settings: A Guide to Qualitative Observation and Analysis*, 3rd Edition. Belmont: Wadsworth Publishing Company.

Munger, F. (ed.) (2002) *Laboring Below the Line: A New Ethnography of Poverty, Low-Wage Work and Survival in the Global Economy*. New York: Russell Sage.

Newman, K.S. (1999) *No Shame in My Game: The Working Poor in the Inner City*. New York: Vintage.

Ong, P., Rickles, J., Matthias, R. and Benjamin, A.E. (2002) California caregivers: final labor market analysis. Paper submitted to the California Employment Development Department.

Parks, J.A. (2003) *No Place Like Home? Feminist Ethics and Home Health Care*. Bloomington: Indiana University Press.

Scheid, T.L. (2004) Our hard jobs are getting harder: consequences of the commodification of care. Presentation to the American Sociological Association, Panel on Carework, August.

Schneider, S. (2003) Victories for home health care workers, *Dollars and Sense*, 249.

Sennett, R. (1998) *Corrosion of Character: The Personal Consequences of Work in the New Capitalism*. New York: W.W. Norton.

Smith, V. (1996) Employee involvement, involved employees: participative work arrangements in a white-collar service occupation, *Social Problems*, 43, 166–79.

Smith, V. (2001) *Crossing the Great Divide: Workers Risk and Opportunity in the New Economy*. Ithaca: Cornell University Press.

Stacey, C.L. (2004) *Love's Labor's Learned: Home Health Workers Caring for Elderly and Disabled Adults*. Unpublished dissertation, UC Davis.

Stone, R.I. (2001) Research on frontline workers in long-term care, *Workforce Issues in a Changing Society*, Spring.

Stone, R.I. and Weiner, J.M. (2001) *Who Will Care for Us? Addressing the Long-Term Care Workforce Crisis*. Urban Institute and Robert Wood Johnson Foundation.

Strauss, A. and Corbin, J. (1990) *Basics of Qualitative Research: Grounded Theory Procedures and Techniques*. Newbury Park, CA: Sage.

Tronto, J.C. (1993) *Moral Boundaries: A Political Argument for an Ethic of Care*. New York: Routledge.

Wardell, M. (1992) Organizations: a bottom-up approach. In Reed, M. and Hughes, M. (eds). *Rethinking Organizations: New Directions in Organizational Theory and Analysis.* London: Sage.

Wharton, A. (1993) The affective consequences of service work: managing emotions on the job, *Work and Occupations*, 20, 205–32.

Wharton, A. (1999) The psychosocial consequences of emotional labor, *The Annals of the American Academy*, 561, 159–76.

Yamada, Y. (2002) Profile of home care aides, nursing home aides, and hospital aides: historical changes and data recommendations, *The Gerontologist*, 42, 199–206.

Chapter 9

Access, boundaries and their effects: legitimate participation in anaesthesia

Dawn Goodwin, Catherine Pope, Maggie Mort and Andrew Smith

The 'crisis' in anaesthesia

Pilnick and Hindmarsh (1999) argue that anaesthetic practice is accomplished by anaesthetic teams in collaboration with the patient and emphasise the necessity to understand how anaesthetic work is created in interactions. This interaction becomes increasingly salient when service reconfiguration and redistribution of work within the anaesthetic team is being considered. In the UK medically-qualified anaesthetists currently provide general and regional anaesthesia for operative procedures, they practice in high dependency and intensive care, provide obstetric epidural services, and acute and chronic pain management services. The 'New Deal for Junior Doctors' restricted the amount of time junior doctors spend training and working in hospitals (Simpson 2004) and, consequently, has significantly reduced the numbers of medical staff available to deliver anaesthetic services. In turn, this has limited the time available for consultant anaesthetists to provide training and supervision and lead commentators to suggest that staffing was near 'crisis point' (Seymour 2004, Simpson 2004). Exacerbating these difficulties (Simpson 2004) is the European Working Time Directive which introduced a 58-hour maximum working week in August 2004 (HMSO 2003).

The Royal College of Anaesthetists have responded to this 'crisis' by attempting to improve the flexibility and accessibility of anaesthetic training, and potentially to reshape the service by delegating some work to non-medically trained specialist practitioners, such as developing a 'respiratory therapist' role in intensive care, and by midwives providing the epidural service. Nurse anaesthetists provide general anaesthesia overseas but development of this role in the UK remains contentious (NHS Modernisation Agency 2003). Policies such as The NHS Plan (Doff 2000), and *Shifting the Balance of Power* (DoH 2002) encourage the development of new roles, service reconfiguration, modernisation and new ways of working. Little, however, is known about how work and knowledge are at present distributed amongst anaesthetic teams and how this may affect attempts to reshape anaesthetic services. The ability of individual practitioners to organise their own activities, and, crucially, those of other professionals, will have implications for how anaesthesia is reorganised. Alongside the redistribution of specific tasks, responsibilities and knowledge, we suggest the circumstances in which practitioners can legitimately initiate action should be considered.

Distributing work across a nursing-medical boundary

Anspach (1987) argues that differentiation between medical and nursing knowledge is linked to the character of the work doctors and nurses perform. Doctors and nurses engage in different sets of daily experiences that define the character of knowledge available to them. Anspach found that physicians in neonatal intensive care had limited contact with patients and relied heavily on diagnostic technology. Nurses, in contrast, relied upon 'interactive' cues or 'gut feelings' they gleaned from continuous contact with infants. The organisation of the neonatal intensive care unit provided an 'ecology of knowledge', a hierarchy structuring different types of work and providing access to different forms of knowledge. According to Anspach, these forms of knowledge do not carry equal weight in decision making; the knowledge that neonatal physicians employed was prioritised by virtue of it being technologically generated and processed, whereas the worth of interactive cues, on which nurses relied, was actively devalued (Anspach 1987: 229).

Svensson (1996: 384) also recognised that 'nurses get to know and observe the patient in an entirely different way from doctors'. However, Svensson argues that this differentiation does not disadvantage nurses; rather, doctors depend on nursing knowledge. Svensson highlights how changes in the organisation of nursing have increased the opportunities for nurses' involvement in decision making, but he indicates that nurses are still wary of encroaching on medical terrain:

> many [*nurses*] emphasise that they are somewhat careful when it comes to the 'purely medical'. This is viewed as intruding upon another's area of competence, and as calling for some caution against presenting oneself in a way that appears challenging (Svensson 1996: 388).

The distribution of work and knowledge becomes increasingly complex as Allen (1997) describes how, when pressed, nurses routinely undertake a range of duties that fall outside their jurisdiction. Nevertheless, where it was possible to avoid additional responsibilities, nurses did so:

> Nurses, however, had not simply incorporated this work into their everyday practice; rather, they undertook informal boundary-blurring work when the doctor was unavailable. When doctors were physically present on the ward, nursing staff adhered to hospital policy and asked the doctor to carry out these tasks (Allen 1997: 511–2).

This shift was not simply a delegation of medical responsibility to nurses; rather it was a response to organisational difficulties and an acknowledgement of the heavy workload doctors faced. Allen points out that such boundary-blurring work often means 'breaking the rules'; nurses working beyond their formal boundaries are unsupported by organisational policies and are potentially exposed to disciplinary action. Allen also found the nurse's confidence and the extent to which s/he trusted individual doctors influenced the distribution of work between doctors and nurses, again highlighting a nurse's vulnerable position.

Tjora (2000) examined the 'boundary-spanning' activities undertaken by nurses in Norwegian emergency communication centres in which nurses handle requests for medical assistance in both emergency and routine cases. When prioritising emergency visits, nurses constructed descriptions of patients in particular terms to influence what the doctor did. When doctors were un-available, nurses would 'regularly draw on professional experience, and pooled knowledge of colleagues to try to diagnose patient's conditions over the phone' (Tjora 2000: 733). As in Allen's study, such 'boundary-spanning' activity occurred only when doctors were absent. Whilst these boundary-spanning act-ivities optimise the use of doctors' time they inhibit nurses from taking owner-ship of 'diagnosis' and underestimate the complexity of nurses' judgements.

These studies indicate that the distribution of work amongst healthcare practitioners is often organised flexibly and tacitly. The character of the work may change subtly depending on who undertakes it, and the work may be re-labelled. This delicate ordering of healthcare work is rarely scrutinised (Hindmarsh and Pilnick 2002: 141). In one of the few studies to look at the accomplishment of teamwork in anaesthesia, Hindmarsh and Pilnick show how anaesthetists and operating department practitioners (ODPs work in a manner comparable to theatre nurses) are simultaneously attentive and respond to one another's actions whilst engaged in seemingly individual tasks:

> Part of learning to be an anaesthetist or an ODP is about developing expertise in reading the embodied conduct of colleagues. The uninitiated do not have an intimate understanding of the potential or likely trajectories of action that will emerge when a colleague has picked up a gas mask, lifts a mask from the face of the patient or approaches with a syringe at particular moments within the anaesthetic room activities. (Hindmarsh and Pilnick 2002: 152).

They observed how talk, ostensibly directed at the patient, served to camouflage collaboration with their colleagues. Hindmarsh and Pilnick direct attention towards the moment-to-moment practices that order and accomplish anaes-thetic work. Such customary practices are implicated in the construction of boundaries between anaesthetists, ODPs and nurses, and the distribution of knowledge as we outline below.

Teamwork differentiated: method and theoretical resources

The data used in this chapter were collected for an ethnographic study of expertise in anaesthesia (financed by the NHS North West R&D Fund, project grant number RDO/28/3/05), between April 2000 and April 2001, in two NHS hospital trusts in the UK. For this study 34 anaesthetic 'sessions' were observed at the primary fieldsite, and five at the second site for comparative purposes. Typically a 'session' involved obtaining consent from anaesthetists, patients and theatre staff, then accompanying an anaesthetist for approximately four hours during an operating theatre list, a morning in intensive care or a period of on-call work. Contemporaneous field notes were taken and transcribed. We observed anaesthetic work in surgical specialties such as general and vascular, trauma and orthopaedics, obstetrics and gynaecology, ear, nose and throat, maxillofacial and dental, ophthalmics, paediatrics and day case surgery, and included practitioners ranging from novice 'trainee' anaesthetists and newly qualified ODPs to consultants, nurses and ODPs of up to 25 years experience. Occasionally, it was possible to 'debrief' with the anaesthetist, sometimes immediately following the observation period, allowing for discussion and clarification on their practice (on one occasion, discussed below, the debrief showed that the observational data had led to an inaccurate assumption), and twice using the observation transcript as an aid for reflection. The observation was primarily undertaken by the first author, a former anaesthetic nurse at the primary field site, and the particular ethical issues that arose are discussed elsewhere (Goodwin *et al.* 2003). We also conducted interviews with 20 staff selected to reflect the range of roles, levels of skill and expertise described above. These were tape recorded and transcribed, lasted between 30 minutes and two hours, and varied in style and purpose, some being quite general and exploratory, and others focused on a critical incident or a recent episode from the interviewee's practice.

In this chapter we are concerned with anaesthetic work in relation to anaesthesia for operative procedures, we do not therefore discuss anaesthetic work in intensive care, pain clinics or other areas. We focus on instances that arose in which the limits of a participant's practice were questioned, with a view to elucidating the processes and patterns through which practice is organised and knowledge is generated. We enlist the concept of 'legitimate peripheral participation' (Lave and Wenger 1991) in 'communities of practice' (Lave and Wenger 1991 and Wenger 1998) to help explore how these boundaries are created and sustained.

A community of practice has three dimensions. In the first – 'mutual engagement' – participants work together contributing to a dense network of working relationships. The second dimension is a 'joint enterprise'; this is the lynchpin holding the community together. This is not a given objective, it is the result of collective negotiation. Thirdly, the community must have a 'shared repertoire' of routines, words, tools, ways of working, stories, gestures, symbols and actions (Wenger 1998). Lave and Wenger's (1991) study of apprentices and Wenger's (1998) later study focused on single occupations. When practitioners from different disciplines form a community of practice the boundaries *within* the community become important, as we will show, rather than those at the margins, which Wenger discusses.

The anaesthetic community of practice, as configured in the operating theatre, consists of an ODP or an anaesthetic nurse, an anaesthetist and a recovery nurse. The ODP/anaesthetic nurse prepares for and receives the patient from the ward into the anaesthetic room, and assists the anaesthetist in the provision of anaesthesia. Although ODPs and nurses have different professional backgrounds, a factor that has engendered considerable dispute (Timmons and Tanner 2004) in our study, nurses with a specialist anaes-thetic qualification and ODPs were considered as interchangeable in a practical sense, for example, on work allocation rotas. We do not, therefore, elaborate on the distinctions between nurses and ODPs. The anaesthetist – a consultant or 'trainee' (so-called as s/he will be enrolled on a seven-year specialist training programme), working either alone or together – assesses the patient, plans and administers the anaesthetic, monitors the patient throughout the surgery, then withdraws anaesthesia and prescribes care for the recovery period. A recovery nurse then takes responsibility for the patient in the immediate post-operative period until the patient can be transferred out of recovery. In addition to these core members, the community of practice may also include medical, nursing and ODP students. This community practises in three main environments – the anaesthetic room, the operating theatre and the recovery room – each environment involving further marginal members; these include ward nurses in the anaesthetic and recovery areas, and surgeons and scrub nurses in theatre.

Lave and Wenger (1991: 35) use 'legitimate peripheral participation' to describe 'engagement in social practice that entails learning as an integral constituent'. For them, knowledge is located within the community, not within individuals, and learning occurs through 'centripetal participa-tion', implying movement towards the centre as participation in practice increases:

> For newcomers, their shifting location as they move centripetally through a complex form of practice creates possibilities for understanding the world as experienced (Lave and Wenger 1991: 122–3).

Whilst the 'centripetal' participation of nurses and ODPs is constrained by the boundaries of their occupation, the term 'legitimate peripheral parti-cipation' remains useful in underscoring these multiple overlapping learning trajectories and the limits to their development. Following Lave and Wenger's argument, that learning is shaped by the distribution of tasks and that attaining legitimacy to practice is more important than formal instruction, we suggest that the acquisition of legitimacy affords a kind of security clear-ance, enabling access to restricted areas, opportunities and experiences. Newcomers need an introduction or a sponsor to legitimate their presence, but like security clearance, legitimacy is *stratified*: the role to which a new-comer aspires correlates with the level of legitimacy, and with the rights to practice that they are granted.

Access, boundaries and their effects

Below, we follow some of the disputes that occurred when the boundaries of a nurse's or ODP's practice were questioned. These struggles illustrate some of the issues Lave and Wenger suggest are integral to learning: the level of access permitted and the degree of legitimate participation. We discuss how limited participation affects the development of knowledge and subsequently the potential a practitioner has to inform anaesthetic practice.

Regulating access: preserving practices
Induction of anaesthesia occurs in the small anaesthetic room adjoining the operating theatre where access is usually limited to those who have a practical function to perform. In the excerpt below a consultant anaesthetist and a trainee, approaching the end of her anaesthetic training, are assisted by a senior ODP, and the 'I' in the data is the researcher. All names are pseudonyms and data editing is indicated by '(. . .)'. Quotes are near verbatim and '. . .' indicates missing fragments. For reasons of expediency the surgeon wants to perform the first procedure, the extraction of two teeth from a child, in the anaesthetic room. The consultant anaesthetist does not object but the ODP does. As we join the scenario the child is lying on a trolley having only just been anaesthetised. As in this case, the application of routine monitoring devices before inducing anaesthesia is sometimes waived for young children, with the monitoring being applied immediately after induction:

> The surgeon enters the anaesthetic room, goes round to the right hand side of the patient (. . .) and then pulls the teeth out. There is an exchange between the surgeon and the ODP that I don't catch. Someone (either the consultant or the ODP) says 'Put the monitoring on!'
> Consultant: 'Done?'
> Surgeon: 'Yes, better find a specimen bottle for these'. He leaves the anaesthetic room (. . .)
> Consultant: 'Pop her on her side'. The child is turned onto her side and the cot-sides of the trolley are raised. He looks at the ODP: 'I'll ask him not to come in the anaesthetic room for the next one'. The consultant is assembling the oxygen mask. The black face mask is replaced with the oxygen mask. The surgeon returns and places the specimen bottle, containing the teeth, under the child's pillow. He turns to the ODP: 'Sorry Steve, it was just . . .'
> The ODP interjects: 'No, my objection was right'. He takes the patient to Recovery with the trainee anaesthetist.

The anaesthetic room is a nexus of anaesthetic knowledge and expertise; it is where monitoring is applied, 'lines' and catheters are introduced, the patient is rendered unconscious, breathing tubes are inserted, and 'nerve blocks' (pain relieving techniques) are performed. The surgeon's encroachment of this territory disrupts routine anaesthetic practices aimed at safeguarding the patient: he removes the teeth before the monitoring is attached. In the recovery room afterwards the consultant anaesthetist went on to concede that the surgeon should be excluded from the anaesthetic room:

> '. . . I have no problem with you (*researcher*) being there but Steve (*ODP*) has already complained about four times this morning that there are too many people in the anaesthetic room . . . he's got a point actually, we should have put the monitoring on first . . .' (. . .) We go back into the anaesthetic room and begin preparations for the next patient.
> Consultant: 'Fentanyl please'.
> ODP: 'That's what I was saying, you can't keep him [*the surgeon*] out'. (whilst opening the controlled drug cupboard).
> Consultant: 'You're right, we should have put the monitoring on first'.

Access to the anaesthetic room permits exposure to anaesthetic room practices and the knowledge suspended within them. Such practices must be observed in order to preserve anaesthetic knowledge, expertise and patient safety. The above incident outlines how knowledge and expertise can be spatially bounded and protected as legitimate access to marginal actors is awarded and revoked. We now look at the boundary disputes between core members of the community which display how work, knowledge and practices are distributed, giving rise to a range of practitioner identities.

Inside the community: legitimacy disputed
Legitimate forms of participation become even more contentious where anaesthetic nurses, ODPs and trainee anaesthetists are concerned. This issue is played out between the ODP and the trainee anaesthetist later the same morning. Explanations of technical terms and devices are in italics:

> As I enter the anaesthetic room the trainee anaesthetist and the ODP are present and the patient is lying on a trolley already anaesthetised. The trainee stands at the head of the patient and the ODP by the patient's shoulders and next to the anaesthetic machine. The trainee removes the laryngeal mask (*LM – a device inserted into the throat of an unconscious patient to hold open the airway and allow for ventilation*). The ODP takes a new face mask out of a packet. I look at the oxygen saturation monitor, it is reading 100 per cent, the patient looks pink, normal colour. (. . .)
> Consultant: 'Is it just a poorly fitting LM?'
> Trainee: 'Umm'.
> ODP: 'It's not down far enough'. (He seems to answer for the trainee.)
> Trainee tries to reinsert the LM, she is unsuccessful.
> ODP: 'Come round this side . . .' (gesturing to the right hand side of the patient).

Trainee: 'I will try it my own way, please, if you don't mind'. She reinserts the LM.

Consultant: 'It's turned, you can tell it's not in right because the black line is twisted'. (*A black line runs along the upper side of a reinforced laryngeal mask.*) Trainee removes the LM.

ODP: 'Come round this side . . .' Trainee follows the ODP's instructions and successfully inserts the LM.

Consultant: 'You've just made Steve (*ODP*) a very happy man. (ODP secures the LM with tape.) You happy?' to the trainee, she nods.

Here, the ODP contests the trainee's expertise attempting to instruct her how to insert the airway device – a simple technique usually mastered early in an anaesthetist's career. This trainee is experienced and approaching the end of her training. She is somewhat resistant to this attempt to instruct her, but when a further attempt fails she successfully follows the ODP's directions. There is a palpable tension that the consultant recognises and which his light-hearted observation about making the ODP a 'happy man' is designed to dissolve. This incident raises questions about the extent to which different community members' participation is legitimated. After the session, the consultant commented on the incident:

It was obvious as soon as we walked in the room the LM wasn't in the right place and I think they had removed it by that time. What had happened was Steve (*ODP*) had put the cannula in and then put the LM in, and it wasn't in right, now it doesn't matter to me who puts it in. . . . Fatima (*trainee*) removed it and was trying to reinsert it, and Steve was trying to tell her how to do it, she said she wanted to do it her way. (. . .) The important thing was that she took it out and tried from the side, the way Steve had suggested and it went in. (. . .) I have no problem with Steve telling people how to do things, it doesn't matter to me who it is.

This consultant is perhaps unusually egalitarian concerning the distribution of tasks. The tension noted earlier reflects the perceived illegitimate participation by the ODP. Wenger (1998) suggests that a learning trajectory defines a member's identity and the consultant's commentary here indicates that the ODP was appropriating experiences central to the development of the anaesthetist's identity. The ODP had apparently cannulated the patient, then incorrectly positioned the LM which the trainee removed. The ODP continues to assert his expertise by answering the consultant's questions and instructing the trainee. (It is perhaps worth noting that the debrief revealed the full sequence of events, something we would have missed if reliant only on the observational data.) Although ODPs may develop anaesthetic knowledge and skills, undertaking such procedures impinges on the rights of a trainee anaesthetist both to develop the skills themselves, and crucially, to perform the techniques and procedures that define their identity. Wenger *et al.* (2002: 146) refer to the potential of communities to *stratify* participants creating distinct classes of members as a 'community disorder'. We use '*stratified legitimacy*' to refer to the extent to which an individual's participation is contingent upon their professional identity, but see this not as something to be remedied (although modification may be desirable) but as a constitutive element in the organisation of healthcare to be engaged and studied.

Fortifying professional boundaries

In contrast to the scene above, the next scenario illustrates a more usual distribution of work and knowledge, and indicates how the stratification of legitimate participation – who has the right to question, to act, in which circumstances – is continually reaffirmed. In an interview, a consultant anaesthetist recalls an incident in which a very ill elderly patient had a cardiac arrest on induction. After giving a combination of drugs to induce anaesthesia and paralysis, the patient went very pale and the team were unable to feel the patient's peripheral pulse:

> We all looked at each other for a couple of seconds and we were all
> saying the same thing: shall we, shan't we? It took maybe 10 seconds
> to establish she hadn't got a carotid pulse either and I felt that we
> probably had to do cardiac massage. (. . .) after the second brief episode
> of cardiac massage (. . .) she just suddenly restored an output, and then
> under the influence of the adrenalin she had a heart rate of 100 and a
> blood pressure 200/100 and she very, very quickly pinked up. (. . .) I said
> 'well let's get on and put the lines in' at which stage I think John (*ODP*)
> and Priya (*trainee anaesthetist*) found it a bit too much because they just
> said 'I don't think you should be going on any further'. That's where
> I think they had a valid point. You could question what on earth they
> are doing saying that in an anaesthetic room in that circumstance, and
> I found that quite challenging actually, particularly from John. I think
> with Priya it's OK, because Priya is in a position of training to make those
> decisions, so Priya has a right to know why I'm doing that. (. . .) John
> has to do what you ask.

Whilst this scene has not been observed the consultant's account does convey his approach to *appropriate* participation. The consultant strongly positions himself as the arbiter of legitimate participation; the trainee anaesthetist had a valid right to question the consultant's decision because her professional trajectory points to *full* membership as a consultant anaesthetist. In contrast, the ODP's questioning was challenging because his professional trajectory reifies his *peripheral* status. The legitimacy of the ODP's participation is restricted to the level of performing prescribed tasks, its 'centripetal' movement constrained. The consultant returns to this issue:

I was actually very angry that he challenged me in the middle of that but in a sense he was right, he was playing it by the book. If we were going to resuscitate this woman we should do it properly, get a few more people along, you know give X mgs of adrenalin, according to a protocol, defibrillate at X joules. But (. . .) this is my patient, only I know what her medical history is, and only I know how difficult it is to resuscitate somebody with aortic stenosis. Therefore, only I am competent to make the decision as to whether or not we progress. I don't need six theatre nurses who have all been on an ALS course telling me what drugs to use!

This interview suggests that whilst nurses and doctors learn alongside one another on 'Advanced Life Support' courses, and must demonstrate the same competencies, in practice the opportunities for participation are once again *stratified* with doctors retaining the interpretive, diagnostic and prescriptive functions, and ODPs and nurses typically performing the prescribed tasks.

The effects of stratified legitimacy on knowledge resources
The above discussion indicated how access and legitimate participation are regulated so as to reaffirm occupational boundaries and support the customary distribution of practices. The scene below illustrates how this stratification of legitimate participation affects the resources a participant has to guide anaesthetic care. A consultant anaesthetist, a trainee, a medical student and the senior ODP (featured in the first scenario) are working together. As we join the scenario the trainee anaesthetist is ventilating the patient and the ODP stands beside him waiting to assist when inserting the laryngeal mask (LM).

ODP is holding the LM.
Trainee: 'I'll just give her a bit more (*anaesthetic*) . . .' he connects the propofol syringe to the cannula and injects. He ventilates then lifts the face mask off. ODP holds the LM hovering over the patient's face.
Trainee: 'No, not yet. (He repositions the guedel airway – *a device to prevent the tongue falling back and occluding the airway.*) Is she biting her tongue? No (quietly)'. He continues ventilating, holding the face mask on with one hand and squeezing the reservoir bag with the other, he repositions the face mask and then resumes ventilating. (. . .) The trainee lifts the patient's jaw and holds the mask on with both hands and looks at the reservoir bag, it moves but is not clearly inflating and deflating. The patient makes muffled groaning noise. (. . .) The trainee turns the Sevoflurane (*anaesthetic gas*) down from eight per cent to five per cent. The reservoir bag is now clearly inflating and deflating.
Consultant: 'So you can see what Peter (*trainee*) is doing, getting her deep and settled so she will accept the LM . . .' (to the medical student).
Trainee lifts the face mask off and suctions, he hesitates but the ODP inserts the LM, it stays in position.

Trainee: 'That's good'. Moments later patient coughs. 'Oh mama mia'.
Consultant: 'This is where 20 a day doesn't help . . . airway irritable. She looks like she's trying to cough the LM out. She is breathing down the anaesthetic so she might actually settle (to the medical student) . . .'

Here the ODP and the anaesthetists evaluate the depth of anaesthesia differently. The ODP relies on 'current' knowledge to interpret the depth of anaesthesia whereas the anaesthetists draw on information from the pre-operative visit – the patient is a smoker and will therefore have an 'irritable airway' necessitating a deeper level of anaesthesia for insertion of the LM. Excluded from the preoperative assessment, the ODP has fewer resources to evaluate the patient's condition and appropriately inform the course of anaesthetic care.

Another example of how the stratification of participation limits the ODP's knowledge resources occurred in the operating theatre. Surgery is already underway and the consultant and ODP stand together looking at the anaesthetic machine.

Consultant mentions the CO_2 (*carbon dioxide measurement*).
ODP: 'What are you thinking? MH?' (*MH – Malignant Hyperthermia, a rare inherited disorder triggered by anaesthetic agents, characterised by climbing temperature and a high carbon dioxide*)
Consultant: 'I'm not really thinking MH, he's had too many anaesthetics, but he shouldn't have a CO_2 of that either'. (. . .)
(*Later, during a quiet period, I ask whether the consultant was worried about the CO_2 measurement.*)
Consultant: 'Yes, because the trace didn't drop to the baseline which means that he will have inspiratory CO_2 which you shouldn't have at all. (. . .) So that means either a leak in the circuit or MH, it's unlikely to be MH as he has had too many previous operations'.

The consultant's problem solving involves the interleaving and verification of many sources of knowledge, but ultimately the patient's medical history mitigates against a diagnosis of malignant hyperthermia, details available to the anaesthetist, but not the ODP.

Initiating action: the persuasive potential of knowledge resources
We now examine the resources ODPs and nurses do have, and their potential to inform the course of anaesthesia. We draw on data set in the recovery room. Of all operating theatre roles, recovery nurses are possibly the most autonomous (Timmons and Tanner 2004). Practising independently, facili-tating the 'emergence' of patients from anaesthesia, the need for recovery nurses to inform the course of events is perhaps more acute, and their

activities therefore more distinct, than for anaesthetic nurses and ODPs who assist doctors. Below a recovery nurse describes how a patient in her care deteriorated following a routine operation. Suspecting an internal haemorrhage the recovery nurse repeatedly raised her concerns with the consultant surgeon and suggested a blood transfusion might be necessary. After briefly examining the patient this suggestion was rejected. The nurse then spoke with the trainee anaesthetist who administered the anaesthetic and obtained a prescription for a blood transfusion. The recovery nurse continues:

> Kate (*the patient*) continued to slowly deteriorate over the course of the morning and at 12.20 I was extremely concerned, agitated and frustrated, having continuously raised my concerns strongly to all involved parties, I felt unsupported and ignored. Kate by this time appeared pale and clammy, her blood pressure was being maintained with colloid (*type of fluid*) infusion but her conscious level was deteriorating and it was becoming difficult to rouse her. (. . .) Again I voiced my concerns, this time with the surgical senior trainee as the consultant had left. During my conversation with the senior trainee Kate's blood results returned from the laboratory. Kate's haemoglobin was recorded at 6.8. It was now obvious that Kate was haemorrhaging internally but by now she was visibly and physiologically shocked. Despite Kate's critical condition the senior trainee remained reluctant to take Kate back to theatre without first discussing it with the consultant surgeon. (. . .) The consultant anaesthetist in charge of the critical care directorate (*entered and*) (. . .) quickly assessed the situation, and began to make immediate arrangements for six units of blood to be transfused as a matter of urgency and for Kate to return to theatre (. . .) whereupon following an emergency laparotomy it was discovered that she had been haemorrhaging from a small incision to her liver.

This incident was not observed. The interview data, however, do highlight how the need for nurses and ODPs to persuade other participants to act orientates their work. The recovery nurse develops an account of the situation but the solution lies outside her remit. She cannot prescribe a blood transfusion or perform surgery so must persuade the doctors to act. Later in the interview the nurse indicated the different resources the nurse and the surgeon draw on:

> The surgeon was so reluctant (. . .) because she didn't display the typical textbook signs of haemorrhage. (. . .) I can remember saying to him 'you need to look at this patient as a whole, look at her holistically, don't look at her vital signs look at *her*, she has deteriorated'. He said 'Well how's she deteriorated, her blood pressure hasn't got any lower?', 'Look at *her*, she was conscious before, she was easily rouseable, she was warm, alright she was pale, but now she's pale, clammy, hypothermic, and I'm having to

put a warming blanket on her, and she's not easily rousable. (. . .) Look
at the patient, don't look at the monitors, look at *her!*'

The lack of 'textbook' indicators strip the nurse of the resources with
which to persuade the surgeons to act. The surgeons rely on measurements,
which are partly being offset by the manipulation of the fluid infusion.
Interestingly, the recovery nurse was able to persuade the anaesthetists. The
work of anaesthetists and recovery nurses are closely aligned, both adjust
their care to the effects of a surgical intervention, they look for the same
signs and indicators and come to appreciate the significance of the less tangible signs. However, the boundaries of her profession dictate that the nurse
must rely on persuasion and cannot initiate action. Legitimacy regarding
diagnosis and prescription is retained within the boundaries of medical
practice.

Regulating participation: processes and effects

Hindmarsh and Pilnick (2002: 158) describe teamwork as 'a practical
accomplishment that emerges despite the fact that team members often have
unequal power or status'. We have focused on the ways in which status, as
configured in identities, is developed and achieved through the stratification
of legitimate participation in anaesthetic practice. We have also elaborated
how a differential in knowledge develops, and the effects this has, in terms
of initiating action, for the nurses and ODPs of an anaesthetic team.
   Lave and Wenger's (1991) concept of legitimate peripheral participation
highlights the processes through which different roles and identities are
sustained. We have demonstrated that legitimate access and participation are
stratified in line with professional or occupational identity. Our departure
from Lave and Wenger (1991) and Wenger (1998) is our focus on a multidisciplinary community. The 'joint enterprise' to provide safe anaesthesia
for operative procedures is partitioned across the members of this community of practice. However, the peripheral positions of the ODP and the nurse
are not transitory resting places for apprentices as they develop a skill
and move on; rather they are designated forms of engagement. One can
engage in anaesthesia as an ODP, a nurse or an anaesthetist but movement
between these positions is not endorsed. Lave and Wenger's emphasis on
legitimacy elucidates members' responses to attempts to cross boundaries –
the exclusion of the surgeon from the anaesthetic room, the friction between
the ODP and the trainee anaesthetist when he appropriated anaesthetic
responsibilities, the consultant's anger as the ODP questioned his judgement
– because movement outside accepted boundaries jeopardises another's
identity.
   Wenger's (1998) discussion of boundaries focuses on those that encompass
the community and that overlap with other communities. Our analysis has

concentrated on the boundaries *within* a community that distinguish and contain the different members and their participation. Legitimate participation in the anaesthetic community of practice involves occupying one of the 'peripheral' positions and performing the tasks consistent with that role. Development here is not a seamless progression towards a central position, there may be some overlap between positions but there are significant constraints on the 'centripetal movement' of nurses and ODPs. Developmental trajectories are contingent upon the adoption of a peripheral position and the degree of legitimate participation awarded in line with that identity. Consequently, the nurses' and ODPs' participation is capped, limiting the resources they have to participate and influence the care of the patient.

The effects of these disciplinary boundaries are felt in terms of initiating action: when the required intervention falls outside a participant's remit, initiating action hinges on persuading other participants to act. Stein's (1967) classic analysis of the doctor-nurse game indicates that this need for persuasion is not new. Indeed, Hughes (1988) describes how nurses do not necessarily cloak their suggestions, but are at times decidedly forthright about their recommendations. Prowse and Allen (2002) suggest that the 'routine' or 'emergency' status of the clinical situation is particularly relevant in shaping nurse-doctor interaction. They describe how nurses adopted a diplomatic and sensitive approach when acting in ways that blurred occupational boundaries in routine situations, whereas in emergencies, status differences were less influential. Our study suggests a need also to consider the orientation of different doctors – anaesthetists and surgeons – and indicates that successfully persuading others to act requires legitimate access to the same resources. Doctors do not draw only on technologically produced measurements, and nurses/ODPs do not only use 'interactive cues', other configurations are possible. The perspectives of anaesthetists and recovery nurse in the final scenario are closely aligned, they interpret the significance of both the technologically-produced measurements and the subtle signs in the same way. This runs counter to Anspach's (1997) suggestion that 'information from diagnostic technology assumes a superior epistemological status' (1987: 219).

Allen (1997) and Tjora (2000) show how nurses undertake work that blurs the boundary between medicine and nursing, particularly when doctors are unavailable, and Tjora contends that nurses do perform diagnostic activities, but do not take *ownership* of it. Our analysis supports this assessment and goes further to illustrate how and why this is problematic. The recovery nurse accurately diagnoses an internal haemorrhage but encounters difficulty when initiating a course of action as those practitioners required to act disagree with her assessment. Lacking legitimacy to prescribe care, her only option is to attempt to persuade another participant whose role formally legitimates the necessary activity. Such persuasion points to how, in order to achieve optimal care of a patient, nurses and ODPs routinely operate beyond the official boundaries of their role. A tension arises when this informal means of organising care fails: officially and unofficially the practitioner has no means of securing appropriate care of the patient.

If the reconfiguration of NHS anaesthetic services follows current trends towards increasing use of protocols and clinical practice guidelines (Timmermans and Berg 2003) the development of non-physician anaesthetists is likely to be a protocol-guided service. This means that whilst the scope of practitioner's work may increase the overall shape of the service, and the relationships between the participants, would remain much the same with doctors retaining their diagnostic and prescriptive capacities, and nurses and ODPs implementing the prescribed care. It is likely that this 'new' arrangement will inherit the same problems as the old, in that boundary 'blurring' will still be necessary but it will remain obscured. When working across boundaries, nurses and ODPs will continue to be unsupported by organisational policies precisely because diagnosis and prescription lie outside the boundaries of their practice. A more radical reconfiguration of the service, to meet the staffing crisis identified in the introduction of this chapter, would require challenging the legitimacy of current boundaries, and re-stratifying participants' access to knowledge and resources that enable them to practice.

Acknowledgements

We would like to thank all the patients and staff who participated in this study and the anonymous reviewers whose comments were most helpful in improving this paper.

References

Allen, D. (1997) The nursing-medical boundary: a negotiated order? *Sociology of Health and Illness*, 19, 4, 498–520.

Anspach, R.R. (1987) Prognostic conflict in life-and-death decisions: the organization as an ecology of knowledge, *Journal of Health and Social Behaviour*, 28, 215–31.

Department of Health (2002) *Shifting the Balance of Power: the Next Steps*. Available from http://dh.gov.uk/assetRoot/04/07/35/54/04073554.pdf [downloaded on 29th June 2004].

Department of Health (2000) *The NHS Plan: a Plan for Investment, a Plan for Reform*. Available from http://www.nhs.uk/nationalplan/nhsplan.pdf [downloaded on 29th June 2004].

Goodwin, D., Pope, C., Mort, M. and Smith, A. (2003) Ethics and ethnography: an experiential account, *Qualitative Health Research*, 13, 4, 567–77.

Hindmarsh, J. and Pilnick, A. (2002) The tacit order of teamwork: collaboration and embodied conduct in anaesthesia, *The Sociological Quarterly*, 43, 2, 139–64.

HMSO (2003) *The Working Time (Amendment) Regulations*. Available from www.hmso.gov.uk/si/si2003/20031684.htm [downloaded on 25th October 2004].

Hughes, D. (1988) When a nurse knows best: some aspects of nurse/doctor inter-action in a casualty department, *Sociology of Health and Illness*, 10, 1, 1–22.

Lave, J. and Wenger, E. (1991) *Situated Learning: Legitimate Peripheral Participation*. Cambridge: Cambridge University Press.

NHS Modernisation Agency (2003) Anaesthesia practitioner trial under way, *New Ways of Working*, Winter 2003, 3–4.

Pilnick, A. and Hindmarsh, J. (1999) 'When you wake up it'll all be over': communication in the anaesthetic room, *Symbolic Interaction*, 22, 4, 345–60.

Prowse, M. and Allen, D. (2002) 'Routine' and 'emergency' in the PACU: the shifting contexts of nurse-doctor interaction. In Allen, D. and Hughes, D. *Nursing and the Division of Labour in Healthcare*. Basingstoke: Palgrave Macmillan.

Seymour, A. (2004) Non-medical delivery of anaesthesia, *Bulletin 24*. Royal College of Anaesthetists.

Simpson, P. (2004) *The Impact of the Implementation of the European Working Time Directive to Junior Doctors Hours on the Provision of Service and Training in Anaesthesia, Critical Care and Pain Management*. Available from http://www.rcoa.ac.uk/docs/ewtd.pdf [downloaded on 29th June 2004].

Stein, L.I. (1967) The doctor-nurse game, *Archives of General Psychiatry*, 16, 699–703.

Svensson, R. (1996) The interplay between doctors and nurses – a negotiated order perspective, *Sociology of Health and Illness*, 18, 3, 379–98.

Timmermans, S. and Berg, M. (2003) *The Gold Standard: the Challenge of Evidence-Based Medicine and Standardization in Health Care*. Philadelphia: Temple University Press.

Timmons, S. and Tanner, J. (2004) A disputed occupational boundary: operating theatre nurses and Operating Department Practitioners, *Sociology of Health and Illness*, 26, 5, 645–66.

Tjora, A.H. (2000) The technological mediation of the nursing-medical boundary, *Sociology of Health and Illness*, 22, 6, 721–41.

Wenger, E. (1998) *Communities of Practice: Learning, Meaning, and Identity*. Cambridge: Cambridge University Press.

Wenger, E., McDermott, R. and Snyder, W.M. (2002) *Cultivating Communities of Practice: a Guide to Managing Knowledge*. Boston, MA: Harvard Business School Press.

Notes on Contributors

Davina Allen is a Professor and Research Director at the School of Nursing and Midwifery Studies, Cardiff University. Her research interests are concerned with the organization and delivery of health and social services and its relationship to clinical effectiveness, service quality and professional education and socialisation. She has published with David Hughes on Nursing and the Division of Labour in Healthcare (2002) and her recent work has focused on the co-ordination of complex care trajectories.

Cecilia Benoit, Ph.D., is Professor in the Department of Sociology at the University of Victoria, Canada, and Senior Research Administrator for the Office of Research Services. Her research interests include comparative welfare states, gender, work, health and marginalization.

Ivy Lynn Bourgeault, Ph.D., is an Associate Professor in Health Studies and Sociology at McMaster University and holds a Tier II Canada Research Chair Comparative Health Labour Policy. Her research interests include health professions, health policy, women's health and complementary and alternative health care.

Yvonne Carter is currently Dean of Warwick Medical School at the University of Warwick. She is also a Professor of General Practice and Primary Care and has a long-standing background in health services research.

Mary Ann Elston's research interests are the organisation and history of health care, particularly in relation to gender and the medical profession, but her recent research projects have included studies of violence against professionals in the community and of the NHS Direct telephone health advice service.

John Germov, Ph.D., is a Senior Lecturer in Sociology in the School of Social Sciences at the University of Newcastle, and is the Immediate Past President of The Australian Sociological Association (TASA). Recent books include: *Second Opinion: An Introduction to Health Sociology* (2005, 3rd ed) and *A Sociology of Food and Nutrition: The Social Appetite* (2004, 2nd ed, with L. Williams). His research interests include food sociology, social determinants of health and management sociology.

Jackie Goode is a Research Fellow in the Institute for Research into Learning and Teaching in Higher Education at the University of Nottingham. Her research interests include the student experience of higher education, problem based learning, health services research and professional development and qualitative methodologies. Recent publications include: Goode, J. *et al.* (2004) 'Male Callers to NHS Direct: The Assertive Carer, The New Dad, and The Reluctant Patient', *Health: an interdisciplinary journal for the social study of health, illness and medicine*, 8, 3, 311–28 and Goode, J. *et al.* (2004) Risk and the Responsible Health Consumer, *Critical Social Policy*, 24, 2, 210–32.

Dawn Goodwin is a Postdoctoral Research Fellow at the Institute for Health Research, Lancaster University. Having completed her PhD 'Acting in Anaesthesia: Agency, Participation and Legitimation' she is currently working on an ESRC/MRC funded project: 'Building Networks of Accountability: Connecting Humans, Machines and Devices'. Her previous publications include 'Ethics and Ethnography: an experiential account', *Qualitative Health Research*, 2004, 13 (4) 567–577 and Expertise in practice: an ethnographic study exploring acquisition and use of knowledge in anaesthesia, *British Journal of Anaesthesia*, (2003) 91 (3) 319–28.

David Greatbatch is a Special Professor in the School of Education at the University of Nottingham, where he is a member of the Centre for Developing and Evaluating Lifelong Learning (CDELL) and an associate of the Institute for Learning and Teaching in Higher Education (IRLTHE). He has previously held positions at the Universities of Oxford, London, Surrey and Warwick, and the Xerox Research Laboratory in Cambridge. He has published in journals such as *American Journal of Sociology, American Sociological Review, Language in Society, Law and Society Review* and *Human Relations.* He is coauthor (with Timothy Clark) of *Management Speak: Why We Listen to What Management Gurus Tell Us* (Routledge) and is currently completing *Agreeing to Disagree: The Dynamics of Third Party Intervention in Family Disputes* (Cambridge University Press, forthcoming).

Gerald Hanlon is a Professor in the Management Centre at the University of Leicester. His research interests include expertise, changing nature of professional labour, the changing relationship between the state, capital and civil society. He is currently leading an ESRC research project on Corporate Social Responsibility.

Donna Luff is a lecturer in the Social Science of Health in the School of Health and Related Research at the University of Sheffield. Her current research interests are in the areas of patient/consumer perspectives in health care, complementary and alternative medicine and gender and sexual health. Her recent publications include articles in *Sociological Review, Psychology Evolution*, and *Gender and Complementary Therapies in Medicine.*

Maggie Mort is Senior Lecturer at the Institute for Health Research and Co-Director of the Centre for Science Studies, Lancaster University. Her publications include: Building the Trident Network: a study of the enrollment of people, knowledge and machines, MIT Press (2002); Remote Doctors and Absent Patients: acting at a distance in telemedicine? in *Science, Technology & Human Values*, 2003, 28(2) 274–295, and 'Telemedicine and Clinical Governance: controlling technology, containing knowledge' in *Governing Medicine: Theory and Practice*, Open University Press, (2004).

Susan F. Murray's current research interests concern international health policy, with a special focus on maternity care; health sector reform and its impact on healthcare organisations, service delivery and experiences of care; and methodologies for complex evaluation.

Alicia O'Cathain is an MRC Fellow and Senior Research Fellow at the Medical Care Research Unit, School of Health and Related Research, University of Sheffield. Her interests include combining qualitative and quantitative methods in the context of Health Services Research, evaluating health services, and eliciting users' views of health services. Recent publications include an exploration of patient empowerment in NHS Direct, published in *Social Science and Medicine*.

Roland Petchey trained initially in social anthropology and subsequently in the sociology of organisations. He has been researching primary care since 1985. He is currently Reader in Health Services Research and Policy at City University, London.

Alison Pilnick is a Senior Lecturer in Sociology in the School of Sociology and Social Policy at the University of Nottingham. Her primary research interests are in the field of professional/client interaction, particularly the way in which changing professional roles are negotiated interactionally, and the impact of new technologies on work organization and practice. She is author of *Genetics and Society* (2002) and a co-editor of the journal *Sociology of Health and Illness*.

Ruth Pinder is Associate Research Fellow at Brunel University and has published widely on chronic illness and disability, ageing, medical education and general practice. She also teaches ethnography on the Master's programme at the Open University. Ruth has recently completed a major ethnography with the Balint Society in general practice, and is currently working on a combined epidemiological/ethnographic study of the role of community forestry in addressing health inequalities.

Catherine Pope is a Senior Research Fellow at the School of Nursing and Midwifery, University of Southampton. Current projects include an evaluation of NHS Treatment Centres and Advanced Access to General Practice. Her research interests focus on healthcare organisation and delivery, and professional practice. Her recent publications include 'Contingency in everyday surgical work', *Sociology of Health and Illness*, 2002, 24, 4, 369–84.

Jane Sandall, Ph.D., is Professor in the Health and Social Care Research Division, King's College, London. Her research interests are reproduction, health policy and health professions.

Sara Shaw, B.A., M.Sc., is Senior Research Fellow in the Department of Primary Care and Population Sciences, University College London. Her background is medical sociology and for the past few years she has worked within academic primary care. She has a particular interest in constructionist approaches to policy analysis and organisational development in healthcare. Her specific area of expertise relates to the use of discourse analysis as a means of exploring the development of primary care research policy, which is the focus of her doctoral work. She led a recent national evaluation of pilot Primary Care Trust Research Management and Governance sites, recently published in *Family Practice* (Shaw *et al.*, 2004).

Andrew Smith's research interests centre on the knowledge base of medical practice, whether systematic reviews of published research evidence or the tacit aspects of professional expertise. He is also interested in risk and safety and their interaction. Recent publications include Making monitoring 'work': human-machine interaction and patient safety in anaesthesia, *Anaesthesia*, 2003, 58, 1070–8.

Clare L. Stacey is a Postdoctoral Fellow at the Institutes for Health Policy Studies and Health and Aging at the University of California, San Francisco. She is currently revising her dissertation into a book on the work identities and labor struggles of home care workers in California. Broad research interests include community based long-term care, low-wage healthcare work and informal/formal caregiving. She is also working with colleagues at UCSF on an ethnographic study of patient stigma in four healthcare organizations.

Tim Strangleman, formerly of the School of Sociology and Social Policy at the University of Nottingham, is senior research fellow and research institute manager at the Working Lives Research Institute, London Metropolitan University. He has written widely on issues of work and employment change. He has carried out research in the railway, coal, engineering and construction industries.

Edwin van Teijlingen, Ph.D., is Reader in the Department of Public Health and the Dugald Baird Centre for Research on Women's Health, University of Aberdeen. His research interests include reproductive health, health promotion evaluation and psychosocial aspects of genetic counseling.

Raymond De Vries, Ph.D., is Professor of Sociology at St. Olaf College, Northfield, USA, and Senior Fellow at the Center for Bioethics, University of Minnesota, Minneapolis. He is currently studying the links between culture and the structure of health care, and working on a social history of the profession of bioethics.

Catherine Will is based at the University of Essex. She has been researching the introduction of statins in the UK, using this as a case study to explore the place and meaning of epidemiological concepts in the clinic and in policy. She is interested in exploring the 'practices' of clinical trials and the relationships that may be built or imagined between research, audit and practice in healthcare.

Sirpa Wrede, D.Soc.Sc., is Academy Research Fellow in the Academy of Finland and researcher in the Swedish School of Social Science at the University of Helsinki, Finland. Her research interests include health care, welfare state, gender and professions.

Index